Book Title: Building Plumbing System Decontamination - First Report on Recommendations

Book Author: Stephen J. Treado; Mark A. Kedzierski; Stephanie S. Watson; Nicos Martys; Kenneth D. Cole;

Book Abstract: This report summarizes the measurement results and recommended procedures for responding to building plumbing system contamination incidents and restoring the water system to safe operation. The recommendations are based on analysis of the results of a measurement and modelling research project investigated contamination and decontamination issues related to building plumbing systems.

Citation: NIST TN - 1652

Keywords: contamination, decontamination, flushing, piping, plumbing, water supply

NIST Technical Note 2009-1652

Building Plumbing System Decontamination- First Report on Recommendations

Stephen Treado
Mark Kedzierski
Stephanie Watson
Nicos Martys
Kenneth Cole

National Institute of
Standards and Technology
U.S. Department of Commerce

NIST Technical Note 2009-1652

Building Plumbing System Decontamination- First Report on Recommendations

Stephen Treado
Mark Kedzierski
Building Environment Division
Building and Fire Research Laboratory

Stephanie Watson
Nicos Martys
Structures and Materials Division
Building and Fire Research Laboratory

Kenneth Cole
Biochemical Science Division
Chemical Science and Technology Laboratory

National Institute of Standard and Technology
Gaithersburg, MD 20899-8530

September 2009

U.S. DEPARTMENT OF COMMERCE
Gary Locke, Secretary

NATIONAL INSTITUTE OF STANDARDS AND TECHNOLOGY
Patrick D. Gallagher, Acting Director

ABSTRACT

This report presents an overview of measurements and analysis of contaminant accumulation and removal in building plumbing systems. In addition, methods for decontaminating building plumbing systems and restoring them to safe operation, based on both specific and generic contaminant characteristics are presented. Measurements consistently showed that most, if not all, of the contaminants did stick to the plumbing material substrates after the initial exposure, while only some (diesel fuel, toluene) showed a substantial reduction from flushing with clean tap water, others required the addition of high levels of chlorine to effect removal (phorate, gasoline, biologicals). Some of the measurements were used to develop fundamental models to predict maximum contamination levels and required flushing times. These models were, in turn, used to develop computer software that can be used as part of a response to a contamination event.

Keywords: contamination, decontamination, flushing, piping, plumbing, water supply

ACKNOWLEDGEMENT

This work was funded by the U.S. Environmental Protection Agency (EPA) under contract #DW-13-92167701-0, with the guidance of project manager Mr. V. Gallardo.

Table of Contents

3

1.0 INTRODUCTION

Adequate supplies of clean, safe drinking water are a prerequisite for buildings and their occupants, including all commercial and residential structures and facilities. In the past, there have been occasional concerns regarding insufficient treatment of water supplies or undesirable migration of contaminants from plumbing system materials into water distribution systems, usually due to some change in environmental or operating conditions. Recently, another issue has risen regarding the potential for accidental or intentional introduction of contaminants into water distribution systems and the need for effective methods for dealing with such events. This report summarizes the measurement results and recommended procedures for responding to building plumbing system contamination incidents and restoring the water system to safe operation. These recommendations are based on analysis of the results of a measurement and modeling research project that investigated contamination and decontamination issues related to building plumbing systems.

The general classes of contaminants that are considered include chemicals such as fuels, solvents, pesticides and poisons, and biological materials, such as bacteria, spores, and toxins (EPA, 2005). Considering the wide range of plumbing system materials, including copper, PVC, iron pipe, solder, rubber gaskets and sealants, many potential combinations of contaminants and substrates are of interest. The presence of chemical deposits and biofilms typically found on the interior of plumbing components present another consideration (Mays, 2000). In addition, plumbing system designs can vary widely; and flow obstructions, water tanks and other water-using appliances can significantly complicate the analysis (Wingender, 2004). As a result, many traditional measurement methods were not sufficient for detecting the presence of accumulated contaminants, and new methods had to be developed and used.

1.1 Background

To reach the objective outlined above, a combination of detailed static and dynamic measurements were done along with a computer simulation. This was aimed at identifying the tendency of various contaminants to accumulate in building plumbing systems. The measurements enable the determination of effective methods for eliminating or rendering innocuous any accumulated contaminants, thereby ensuring restoration of a safe water supply. The basic measurement methodology involved exposing a particular plumbing material substrate, component, or system to a water/contaminant mixture followed by a flushing or other decontamination activity. This occurred while periodically monitoring or collecting samples of the substrate and/or water to evaluate for the presence of the contaminant or any residual materials. A comprehensive description of these measurements has been published previously by Treado (2005, 2006), and Treado et al. (2006), but is also summarized in the following sections. Testing conditions included both static and dynamic configurations, and ranged from small samples of substrate materials to full-scale building plumbing systems.

Figure 1.1 presents a project flowchart showing the various project tasks and their relationships and general information flow. Two key points which should be noted are the iterative process of the development of new measurement methods based on the results from initial measurements and the incorporation of detailed simulations of contaminant transport and dispersal to help generalize the measurement results. All of the measurement and modeling results fed into the development of decontamination procedures and recommendations in order to make the recommendations as robust as possible.

Numerous contaminant exposure tests were conducted using various combinations of contaminants and substrates under different flow conditions and configurations, including coupons, small pipe sections, full-scale pipe loops, and hot water heater tanks. The range of test conditions was intended to cover what might be encountered in an actual contamination event, although conditions beyond the range might certainly be possible in extreme or unusual cases.

The reason that different measurement methods and configurations were used was that each task focused on a different aspect related to the mechanisms that are responsible for contaminant accumulation and removal. For example, the static bench-top tests concentrated on measuring the fundamental interactions between waterborne contaminants and plumbing system materials, while the bench-top biological tests introduced the effects of flow and biofilms in a controlled manner. The dynamic flow fluorescent measurements allowed in situ measurement of the thickness of the layer of adsorbed diesel fuel under controlled flow conditions. However, the screening and full-scale tests provided measurement data for more realistic plumbing system configurations but under less well-controlled testing conditions.

The contaminants tested included:

- Diesel fuel;

- Gasoline;

- Toluene;

- Strychnine;

- Cyanide;

- Phorate;

- Mercuric Chloride;

- *E. coli;*

- *Bacillus anthracis or thuringensis (BA, BT);* and

- Ricin;

The material substrates tested were:

- Copper;

- Galvanized iron;

- PVC or CPVC;

- Rubber; and

- Brass.

Table 1.1 summarizes which contaminants and substrates were tested for the various tasks of the project. In addition, substrate tests were conducted using previously used hot water heater tanks with diesel fuel, cyanide, strychnine, and BT spores.

Table 1.1		Summary of Contaminant and Substrate Tests			
Substrate	**Copper**	**Iron**	**PVC or CPVC**	**Rubber**	**Brass**
Diesel fuel	A, C, D, E	A, C	A, C, E	A	A
Gasoline	A, C	A, C	A, C	A, C	A
Toluene	A	A	A	A	A
Strychnine	A, C, E	A, C	A, C, E	A, C	A
Cyanide	A, C, E	A, C	A, C, E	A, C	A
Phorate	A, C	A	A	A	A
Mercuric Chloride	A	A	A	A	A
E. Coli	B		B		
BA or BT	B, E		B, E		
Ricin	B		B		

Legend:
A-   Bench-top Static Tests
B-   Bench-top Biological Tests
C-   Dynamic Flow Fluorescence Measurements
D-   Screening tests
E-   Full-Scale Dynamic Tests

In addition to the measurements of contaminant behavior in building plumbing systems, the impact of plumbing system design and operation on decontamination strategies was investigated. Because building plumbing systems are typically composed of a complicated network of water supply piping, fittings, valves, and fixtures, along with a sanitary drain system and water-using appliances, the potential for contaminant accumulation and the associated strategies for removal require careful consideration of real-world factors.

1.2 Potential Contamination Scenarios for Building Plumbing Systems

In general, contaminants may enter a building water supply anywhere upstream, as represented generically by the four locations shown in Fig. 1.2 and summarized below:

1. Contaminants that are introduced far upstream from the building, either before or near the water treatment facility, and travel a significant distance through the water distribution system to reach the building service line;
2. Contaminants that are introduced into a water main supplying multiple branch lines, including one that supplies the building;
3. Contaminants that are introduced near to, but outside of, the building via the building service line; and
4. Contaminants that are introduced to the building water supply from within the building.

The reason that these scenarios are differentiated considers the fact that the methods and the amount of contaminant introduced would probably vary. As a result, the duration and level of contaminant concentrations in the water supply system would vary. The portions of the water distribution system, including the building plumbing system, that are affected would be different. Thus, the response to the contamination events, including methods to control the spread and the procedures to remove the contaminant would differ depending on where the contaminant was introduced.

In the first scenario, the effects of dilution and water treatment will likely reduce, but not necessarily eliminate, the impact of most contaminants on downstream building plumbing systems. The closer the point of contaminant introduction is to the building, the higher the concentration and the greater the potential for contaminant accumulation. However, longer lengths of affected water supply lines will require longer flushing times.

Most contamination events will be recognized based on a consumer complaint (odor, color, and/or taste), illness or reaction, a sensor reading (unlikely), or a verbal or written threat. The first step would be to determine if there is an actual contaminant present in the water supply, what it is, and the extent of it. The point of contaminant introduction can be deduced by collecting water samples from a range of locations. Then map the water lines that are found to be contaminated, working upstream until unaffected water lines are found. In Fig. 1.2, Point 3 has been determined to be the point of contaminant injection. Once the affected water lines are identified, several concerns follow:

- Precisely determining how long the contaminant has been in the water supply system or how much of the contaminant was introduced will be difficult;
- Collected water samples can identify the contaminant, but the sample contaminant concentration will likely be less than that at the point of injection;
- Contaminant may have accumulated in the water supply system, so simply flushing out the contaminated water may not remove all of the contaminant; and
- Additional flushing or special procedures may be needed to restore a safe water supply.

Based on these concerns, assume a worst case exposure condition, which would be a highly concentrated contaminant introduced into the water supply system and interacted with substrates in contact with the water. Those contaminants that accumulate on plumbing system surfaces exposed to contaminated water will need to be eliminated before the system can be restored to safe operation and use. The accumulation may be due to the combined effects of a number of mechanisms, including different types of adsorption, chemical reactions, and sedimentation on/with pipe materials and pipe scale. It may be possible to remove some small sections of piping or other components for laboratory analysis, but that is not always practical or possible.

The magnitude and the location of contaminant accumulation will be a function of the characteristics of the contaminant/substrate interactions and the exposure conditions. For example, some substrates are more conducive to contaminant accumulation. Longer exposure times and higher concentrations can lead to greater accumulations. Contaminants may or may not be soluble in water and more or less dense than water. Soluble contaminants will dissolve, and mix with water to come into contact with most of the plumbing system surfaces. Insoluble contaminants will either float (specific gravity, SG<1) or (sink, SG>1) thereby preferentially coming into contact with the upper or lower inside surfaces of pipes and tanks. See Figs.1.3a and 1.3b. Contaminants with sedimentary characteristics, such as bacteria and spores, may sink due to gravity when water is not flowing, and collect at the bottom or top of plumbing system components. See Figs.1.3c and 1.3d.

The above considerations imply that the particular details of the contaminant characteristics and the method of introduction to a plumbing system can strongly influence the accumulation of the contaminant within the system. In most cases it will be difficult to reconstruct the precise details of such an introduction; therefore, presume the worst-case assumption. This assures safety.

The following section summarizes the measurement procedures and methods used in the analysis. The next section describes the new methods that were developed to reach project objectives.

2.0 PROCEDURES AND METHODS
2.0 Bench Scale Tests
This section provides the procedures and methods used to measure chemical and biological contaminants on the bench scale.

2.1.1 Chemical Contaminant
2.1.1.1 Static Adsorption and Decontamination Experimental Procedures
Interaction of the chemical contaminants in water with pipe materials was investigated by monitoring the changes in the concentration of the specific chemical contaminant in water over time. Experiments took place in a 600 mL beaker (500 mL of contaminant solution) or 500 mL capped jar (450 mL contaminant solution) for volatile organic compounds and solutions were stirred with a magnetic stir bar to accelerate equilibration time. Pipe materials consisted of copper (Cu) pipe, and polyvinyl chloride polymer

(PVC) pipe. Pipe scale consisted of calcium carbonate ($CaCO_3$) powder to represent lime scale. Clean (new) Cu pipe flat coupons were also prepared from new Cu pipe that was sectioned and flattened. All freshly cut new pipe materials were cleaned in methanol to remove any cutting oils. Samples of Cu pipe (used as a line from a hot water heater) from a local apartment building were also obtained and labeled as Cu in-service pipe. Biofilm was grown on a selection of new Cu and PVC pipe materials. Biofilms were grown on clean pieces of Cu and PVC tubing that were sectioned then cut in half and held together with shrink-wrap tubing. Biofilm was grown on the pipe sections by flowing synthetic water plus humic acids, an energy source for bacteria, for a period of two weeks and then for three days with only synthetic water. Pipe pieces with biofilm growth were carefully removed from the shrink-wrap tubing and characterized using FT-IR microscopy before starting adsorption experiments. For each contaminant concentration, two replicates of each pipe material were completed.

A range of chemical contaminant concentrations was chosen based on lethal dosage data and chemical solubility (Budavari, 1996). Initial adsorption studies used deionized water as the solvent, and later adsorption experiments used laboratory tap water for the solvent for more realistic conditions. Due to the immiscibility and the greater density of some chemical contaminants compared to water and to ensure a well-mixed solution, mixing of chemicals in water took place in two increments: one-half the amount of water with the entire chemical contaminant quantity for 20 min followed by mixing the remaining half of water for 20 min. An aliquot of contaminant solution was taken to determine the concentration of the initial chemical contaminant. After pipe material addition, aliquots of the contaminant solution were taken over time periods to monitor the change in chemical contaminant concentration with pipe exposure time. Aliquots of contaminant solution were taken over a length of 5 days using a graduated syringe. With pipe material addition, the adsorption process began and subsequent contaminant solution specimens were taken at 10 min, 20 min, 40 min, 60 min, 120 min, 240 min, and 360 min after the initial pipe sample introduction. Contaminant solution specimens were then taken every morning and afternoon for the following 4 days. Concentration profiles of chemical contaminants in water with the pipe material were then produced. Exposed pipe materials were examined for residual chemical contaminant to determine chemical contaminant surface adsorption after the experiment was complete.

Decontamination experiments consisted of mixing a series of pipe materials that were previously exposed to a specific chemical contaminant at a certain contaminant concentration with a decontaminating water solution. Decontaminating water solutions consisted of fresh tap water, tap water with various bleach concentrations, and tap water with a specific concentration of detergent additive, Spic and Span (SS) detergent[1]. For 450 mL of tap water, bleach concentrations ranged from 21.1 mL, 10.6 mL, 5.3 mL, to 2.6 mL. The concentration of SS used in the decontamination studies was 7.1 mL detergent in 450 mL tap water. During decontamination, both decontaminant water

---

[1]Certain commercial equipment, instruments, or materials are identified in this report in order to specify the experimental procedure adequately. Such identification is not intended to imply recommendation or endorsement by the National Institute of Standards and Technology, nor is it intended to imply that the materials or equipment identified are necessarily the best available for the purpose.

solution and pipe material were examined for evidence of decontamination, i.e., contaminant chemical presence in the decontaminating water solution and a reduction or removal of contaminant on the pipe material. Decontaminant water solution was monitored over an extended time interval during mixing with contaminated pipe material. Pipe material was examined before and after the decontamination experiment.

Decontamination of cyanide exposed pipe materials was performed in a beaker in which 500 mL of decontamination water solution was mixed with a stir bar and with direct, constant monitoring using electrodes. Contaminated pipe samples were added to the decontamination water after initial readings of pH and cyanide concentration using electrodes. Cyanide (CN⁻) concentration and pH were recorded on a 30 min interval for 3 h. When no change in CN⁻ concentration after this decontamination time was observed, the sample was removed from the decontamination water and the decontamination process was considered complete. If a change (or increase) was noted in the concentration, the test continued until the concentration stabilized. If necessary, the decontamination was repeated with another 500 mL of fresh decontamination water.

Decontamination of organic contaminant exposed pipe materials was performed in a 500 mL capped glass jar. Contaminated pipe samples were placed in 450 mL decontamination water and mixed with a stir bar. On an hourly interval, a 10 mL or 40 mL aliquot of the water was collected for water analysis and the pH was also recorded. Deionized water was used to dilute to 40 mL. The final dilution was analyzed and two replicates were analyzed.

2.1.1.2 Conventional Water Analysis

The quality of the water used in the adsorption and decontamination experiments was tested by using a series of standard tests to check the drinking water quality commonly employed within the water supply infrastructure. The tests included pH, conductivity, chlorine content, alkalinity, turbidity, and total organic carbon content (TOC). The standard water tests were used to monitor the water quality before and after both adsorption and decontamination methods to determine changes.

The pH was measured using a combination pH electrode (Accumet, Thermo Fisher) on a electrode meter with ion selective electrode (ISE), conductivity, and computer interface capabilities (Accumet Research AR50 meter, Thermo Fisher) (AWWA, 1990). The pH electrode was calibrated daily with two buffer solutions ranging from pH 4.00, 7.00, and 11.00 depending on the range of pH to be measured in the test solution.

Conductivity was measured on a conductivity meter (Accumet Basic AB30 conductivity meter, Thermo Fisher) using a glass 2-cell, 1.0 cm⁻¹ conductivity cell (Accumet glass, Thermo Fisher). The conductivity cell was calibrated daily with a 0.01 mole/L potassium chloride solution with a known conductivity of 1408.8 µS/cm (ASTM, 1999).

Chlorine content (free and total) was measured using a portable photometer (Pocket Photometer, HF Scientific). The meter was zeroed daily according to manufacturer's instructions and checked against chlorine solutions with a known chlorine concentration, which was prepared according to ASTM Method D512-89 (ASTM, 1999). In the portable photometer procedure, solid reagents containing a chlorine-indicating dye were added to a water sample in a cuvette. Proper mixing of the reagent with the water solution and the use of the defined time for color development was necessary for consistent results.

Alkalinity was measured using a portable water analysis system (Mini-Analyst Series 942, Orbeco-Hellige). The meter was zeroed daily according to manufacturer instructions and checked against a sodium carbonate solution having a 160 mg/L alkalinity value (AWWA, 1991a). In the portable system, a solid reagent was added to a water sample in the sample tube. Proper mixing of the reagent with the water solution was necessary for consistent results.

Turbidity was measured using a turbidity meter (LaMotte 2020). The meter was calibrated daily using turbidity standards in the range of 1.0 NTU to 10.0 NTU (AMCO Primary turbidity standard, LaMotte). In addition, turbidity reference solutions were also prepared according to EPA Method 180.1 (EPA, 1978).

Total organic carbon (TOC) was measured using an automated TOC analyzer (Phoenix 8000 UV-Persulfate TOC Analyzer, Teledyne Tekmar). TOC measurement involves oxidizing organic carbon in the sample to produce carbon dioxide (CO_2), detecting and quantifying the oxidized carbon or CO_2, and presenting the results in units of mass of carbon per volume of sample. The TOC analyzer used a wet chemical method with persulfate oxidation along with UV irradiation simultaneously and nitrogen to sweep the resulting CO_2 to the detector, a nondispersive infrared (NDIR) detector (AWWA, 1991b). Forty mL autosampler vials were used and reagents were 10 % volume fraction persulfate in 5 % volume fraction phosphoric acid and 21 % volume fraction phosphoric acid (from sodium persulfate and concentrated phosphoric acid, respectively). The analyzer reagents lines were primed and cleaned daily. The analyzer was calibrated weekly using a 1000 mg/L carbon standard solution (Teledyne Tekmar) diluted between 0.1 mg/L to 20 mg/L. Two aliquots were analyzed from each autosampler vial.

2.1.1.3 Phorate Water Analysis

Phorate concentrations used in adsorption studies were 12.4 mg/L, 24.8 mg/L, 100 mg/L, and 400 mg/L. Contaminated water was analyzed with an automated purge and trap-gas chromatograph/mass spectral detector (PT-GC/MS) (Teledyne Tekmar XPT and Thermo Finnigan Trace GC-DSQ MS) (EPA, 1995a). The GC method used was as follows: GC column: HP-Ultra2 (fused silica column cross-linked with 5 % phenyl methyl silicone), column length: 25 m, column ID: 0.22 mm, and film thickness: 0.33 μm.

Oven Parameters:
- Initial Temperature: 40 °C for 2 min

- Ramp 1, 14 °C/min to 85 °C and hold for 2 min

- Ramp 2, 30 °C/min to 220 °C and hold 2 min

- Ramp 3, 13 °C/min to 260 °C and hold for 1 min

GC Conditions:
- Inlet Temperature: 250 °C

- Constant Flow at 1.0 mL/min

- Split Inlet with a Split Ratio of 10

The PT conditions were based on default settings with some minor adjustments for longer rinse and purge times between samples to avoid interference. In general, the critical PT parameters used were:

Purge:
- Valve Oven Temperature: 150 °C

- Transfer Line Temperature: 150 °C

- Purge Time: 11.0 min at ambient temperature

- Purge Flow: 40 mL/min

Desorb:
- GC begins at desorb start

- Desorb Preheat Temperature: 245 °C

- Desorb Time: 2 min

- Desorb Temperature: 250 °C

- Desorb Flow: 300 mL/min

Bake:
- Bake Temperature: 270 °C

- Bake Flow: 400 mL/min

- Bake Time: 2 min

A 10 mL aliquot was taken to ensure enough solution remained after the entire experiment for further analysis. The contaminant solution was diluted with deionized water to the 40 mL GC autosampler vial capacity. After dilution, the GC autosampler vials were inverted 5 times to allow for complete mixing. Two GC analyses were completed with each autosampler vial.

The GC peaks at retention time of 12.63 min [$C_7H_{17}O_2PS_3$; phorate] and 8.69 min ($C_5H_{12}S_2$; 1,1' (methylenebisthio)bis-ethane] were monitored to follow the change in phorate concentration.

2.1.1.4 Toluene Water Analysis

Toluene concentrations used in adsorption studies were 500 mg/L, 1.9 g/L, 3.8 g/L, 7.5 g/L, 15.1 g/L, and 30 g/L. The PT-GC/MS method was used to monitor the change in toluene concentration with pipe exposure (ASTM, 1995; EPA, 1989a, 1989b, 1995b). The same GC column (HP-Ultra2) was used as in the phorate experiments with some modifications to the GC method:

Oven Parameters:

- Initial Temperature: 40 °C for 2 min

- Ramp 1, 14 °C/min to 85 °C and hold for 2 min

- Ramp 2, 30 °C/min to 220 °C and hold 1 min

GC Conditions:

- Inlet Temperature: 250 °C

- Constant Flow at 1.0 mL/min

- Split Inlet with a Split Ratio of 10

The PT parameters were the same as that used in the phorate experiments. The GC peak at retention time of 4.0 min (C_7H_8; toluene) was used to monitor the change in toluene concentration.

2.1.1.5 Gasoline Water Analysis

Gasoline concentrations used in adsorption studies were 100 mg/L, 300 mg/L, 500 mg/L, 1000 mg/L, and 2000 mg/L. The PT-GC/MS method that was used for toluene was used to monitor the change in gasoline with pipe exposure (ASTM, 1995; EPA, 1989a, 1989b, 1995b). The GC peaks at retention times of 2.52 min (C_6H_6; benzene), 3.82 min (C_7H_8; toluene), 5.12 min ($C_6H_5C_2H_5$; ethylbenzene), 6.67 min ($C_6H_5C_3H_7$; propylbenzene), and 7.52 min (1,2,3 $(CH_3)_3C_6H_3$; 1,2,3 trimethylbenzene) were used to monitor the change in gasoline concentration.

2.1.1.6 Diesel Fuel Water Analysis

Diesel fuel concentrations used in adsorption studies were 100 mg/L, 300 mg/L, 500 mg/L, 1000 mg/L, and 2000 mg/L. The PT-GC/MS method that was used for toluene was used to monitor the change in diesel fuel with pipe exposure (ASTM, 1995; EPA, 1989a, 1989b, 1995b). The GC peaks at retention times of 7.67 min (1,2,3-$(CH_3)_3C_6H_3$; 1,2,3 trimethylbenzene), 9.45 min (2,4-$(CH_3)_2C_6H_3CH=CH_2$; 2,4 dimethylstyrene), 9.54 min ($C_{10}H_{12}$; 1,2,3,4 tetrahydronapthalene), and 9.93 ($C_{11}H_{14}$; 1,2,3,4 tetrahydro-2 methylnapthalene) were used to monitor the change in diesel fuel concentration.

2.1.1.7 Strychnine Water Analysis

Strychnine concentrations determined for adsorption studies were 0.24 mg/L, 0.47 mg/L, 0.97 mg/L, 1.91 mg/L, and 4.01 mg/L. A method using high pressure liquid chromatography (HPLC) was designed and based on the work of Alliot et al. (1982) which used a chloroform extraction procedure at basic pH (5 mole/L NaOH) with a 10 µg/mL quinine internal standard, all reconstituted in methanol. The HPLC components in this study consisted of a Waters Breeze HPLC system (Waters 1525 Binary Pump, Waters 2487 Dual λ Absorbance Detector, Waters 717Plus Autosampler) with a C_{18} column [5 µm, 2.1 mm x 150 mm (Sunfire, Waters)] and a concentrated ammonium hydroxide and methanol (0.75 : 99.25 volume fraction) mobile phase using a 10 µL sample injection.

2.1.1.8 Cyanide Salts Water Analysis

Cyanide salts used in this study consisted of sodium cyanide and potassium cyanide in concentrations of 3 mg/L, 10 mg/L, 20 mg/L, and 50 mg/L. The cyanide ion (CN^-) was detected using an ion selective electrode (ISE) (CN^- 9606 combination electrode, Thermo Orion). The ISE sensor was cleaned daily with emery paper to remove deposits. The CN^- ISE could not be exposed to CN^- concentrations greater than 25 mg/L due to severe sensor erosion. The CN^- ISE was calibrated daily using CN^- standard solutions (2 mg/L and 20 mg/L from sodium cyanide or potassium cyanide depending on the experiment) prepared according to EPA Method 9213 (1996) and the appropriate ionic strength adjustor (ISA) (1 mL/100 mL solution of 10 mole/L sodium hydroxide). An in-house computer program interfaced to the electrode meter (Accumet Research AR50 meter, Thermo Fisher) was used to constantly monitor the pH, temperature, and CN^- concentration over the entire pipe exposure period.

2.1.1.9 Mercuric Chloride Water Analysis

Mercuric chloride concentrations used in adsorption studies were 20 mg/L, 50 mg/L, 100 mg/L, 300 mg/L, and 500 mg/L. In this study an automated cold vapor atomic adsorption spectrometer (CVAAS) (Hydra C, Teledyne Leeman Labs) was used to measure mercury content in both experimental solutions and pipe materials. Calibration standards were prepared from a 100 µg/L Hg standard (Teledyne Leeman) diluting with 5 % volume fraction nitric acid (15). The instrument was calibrated using Hg standard solutions ranging from 0.1 µg/L Hg to 1.0 µg/L Hg. All liquid samples were diluted to 5 % volume fraction nitric acid to ensure Hg dissolution. Nickel boats, which hold all specimens, were initially heated to 700 °C to remove trace Hg and then periodically heated under the same conditions to clean sample contamination. Sample aliquots ranged from 50 µL to 300 µL, noting that larger aliquots took longer to dry. Solid samples were run as received.

2.1.2 Biological Contaminants
2.1.2.1 Microbiological Measurements of Biofilms, Bacteria, and Spores

An important first step in simulating a water system for testing was establishing a realistic biofilm on the surface of the coupons or pipe sections. The approach was to use a synthetic formulation of tap water supplemented with humic acids (Morrow et al., 2008).

The humic acids stimulated the naturally occurring water microorganisms to grow a biofilm in a short period of time (approximately 3 weeks). This allowed a sufficient number of experiments to be completed within a reasonable amount of time. The biofilm bacteria were removed from pipe or coupon surfaces by vigorous scrapping with a sterile plastic cell scraper. Rinsing the surface with phosphate buffered saline (PBS, 0.01 M phosphate, 0.138 M NaCl, 0.0027 M KCl, pH 7.4) was completed, and the biofilm organisms were dispersed by vortexing for 30 s (Morrow et al., 2008). The solutions were diluted in PBS, plated on R2A nutrient media at ambient temperature, and monitored for growth of up to a week (Schwartz et al., 2003).

The bench scale testing on biological contaminants was done by using simulants for the dangerous biological threats. The *Bacillus thuringiensis* (BT) spore solutions used in the experiments were a commercial preparation (Thuricide, Bonide Insecticide, Oriskany, NY) prepared as previously described by Morrow et al. (2008). The choice of BT as a simulant for *B. anthracis* (BA) spores is based on the close genetic relationship (Radnedge et al., 2003) and similar structure of their outer most layer (the exosporium) (Matz et al., 2001). BT spores can be used at Biological Safety Level 1 (BSL-1) and are suitable for large-scale experiments where it is difficult to maintain higher levels of biological safety. BA (Sterne strain) spores, also known as the vaccine strain were used because they lack the genetic element (pX02 plasmid) needed for virulence in humans and animals. The BA (Sterne) spores require BSL- 2 laboratories and were used for limited experiments in laboratories of the Biochemical Science Division. The properties of the BA (Sterne) spore samples were previously described by Almeida et al. (2006, 2008). The concentrations of the spore samples were determined by the spread plate method. The samples were diluted with PBS buffer containing 0.1 % (volume fraction) Triton X-100. It was essential to include the detergent Triton X-100 to avoid the spores sticking to the plastic tubes used for dilutions. The dilutions were first platted on Luria-Bertani (LB) nutrient agar and then incubated at 35 °C overnight before counting the colonies.

2.1.2.2 *Measurement of Inactivation of Bacteria in Solution*

The inactivation kinetics of the spores and bacteria were measured in solution by contacting the samples with solutions of either active chlorine or monochloramine. Free chlorine concentrations were measured using N, N-diethyl-p-phenylenediamine reagent and chlorine standards (Hatch Company, Loveland, CO). A calibrate curve was measured using chlorine standards (Hatch Company, Loveland, CO). Free chlorine solutions were made by dilution of commercial bleach (sodium hypochlorite). The chlorine concentration was measured after 30 min of stirring. Monochloramine formulations were prepared (Camper et al., 2003) and measured using the indophenol method and standards (Hatch Company, Loveland, CO).

The spores or bacteria were added to either a buffer or synthetic water formulation at a concentration of approximately 10^6 colony forming units (CFU)/mL. A small Teflon-coated stir bar was added and the solution gently stirred to keep the bacteria or spores from settling. The inactivation studies for BT and BA spores were done in small sterile glass vials to prevent the loss of spores by sticking to the walls of plastic containers. The spores tended to stick to plastic tubes unless a detergent such as Triton X-100 was added. Samples of the solution were taken at time zero and at time intervals. Thiosulfate was added to a concentration of 7.5 mM to stop the activity of the bleach or monochloramine. The samples were then diluted with PBS containing 0.1 % Triton X-100 and the dilutions were plated on LB plates to determine the viable spores remaining.

2.1.2.3 Measurement of Ricin Activity

Ricin is a protein toxin found in high concentrations in castor beads. The protein consists of two subunits. The B subunit is responsible for binding ricin to mammalian cells and the A subunit is an N-glycosidase enzyme that catalytically inactivates ribosomes once it reaches the interior of the cell (Endo et al., 1987; Doan, 2004). Stopping protein synthesis results in cell death (cytotoxicity). The measurement of the biological activity of ricin therefore requires the use of mammalian cells grown in culture. Vero cells (CCL 81, ATCC, Manassas, VA, U.S.A.) that were seeded at 6×10^3 cells per well of tissue culture treated 96 well microtiter plates overnight at 37 °C in 5 % CO_2 (Cole et al., 2008) were used. The ricin samples were diluted in cell culture media containing serum and added to the Vero cells. After 22 h the cytotoxic effects of ricin were measured using the yellow tetrazolium MTT assay (#30-1010K, ATCC, Manassas, VA U.S.A.) (Cole et al., 2008).

3.0 NEW METHOD DEVELOPMENT
3.1 Bench Scale Tests
3.1.1 Chemical Contaminants
To simplify conditions involved in pipe systems for accurate adsorption isotherms to be measured, a series of static adsorption experiments were performed in beakers with a watch glass placed on top or in screw-capped jars. Basically, the experiments consisted of adsorption experiments of test aqueous solutions containing a contaminant with a substrate phase representing the pipe material and/or a water deposit followed by a desorption phase. The substrates and the solutions were periodically sampled to determine contaminant accumulation and removal rate. The physical microstructure of the substrate was characterized to aid in the modeling and interpretation of the sorption processes.

Due to the large number of variables to be studied within a pipe system (different grades of pipe, mixtures of deposits in the pipes, temperature, and pH of water, etc), initial adsorption studies were limited to one temperature (ambient), pH of tap water (near neutral), copper and PVC pipes, calcium carbonate, and varying concentrations of contaminants. The solutions were stirred with a magnetic stir bar to accelerate equilibration time, to prevent clumping of the $CaCO_3$ and to ensure equal contact with volatile contaminants. Glass stir bars were used to prevent wear observed in Teflon stir bars and avoid contamination (organics with Teflon). Deionized water was also used initially to simplify interactions and establish analytical methods. All adsorption experiments used laboratory tap water, which was characterized daily, as the solvent. Pieces of pipe were used as specimens or substrates because the interior of whole or parts of pipes were difficult to analyze after adsorption experiments without cutting or penetrating the pipe and risking the disturbance of the adsorption process. Substrate samples of approximately 100 mm^2 surface area were placed on the bottom of the beaker. Suspending pipe samples using a wire mesh basket did not work as the metal wire reacted with contaminant. Experiments were done to determine the solubility of CaCO$_3$ in water; the mass loss observed ranged from 2.2 % to 3.5 %. A known, constant volume (500 mL or 450 mL) of contaminant/water was used for all measurements. The volume was large enough to allow for the removal of a series of aliquots for contaminant concentration analysis over the adsorption time. The dimensions (particularly surface area) and/or mass of pipe substrate were recorded. Four replicates for each contaminant/pipe materials combination were performed. Analysis on the various pipe materials and contaminant concentrations were completed within the project deadlines which established an accurate deviation value. Both the inner and outer surface of the pipe materials were exposed to the contaminant/water solution. Only one side (inner surface) of the exposed pipe material was analyzed for adsorbed species and reaction products. The contaminant/water solution was also monitored for changes in the concentration of contaminant over the adsorption time. Automated methods of water analysis were tried to help facilitate the large volume of specimens. For cyanide experiments using ion selective electrodes, the addition of ionic strength adjustor reagent was omitted to prevent interactions with the chemical contaminant and maintain realistic pH changes. Several analytical methods were tested, but many of the methods, specifically solid phase microextraction (SPME), were found to lack the quantification necessary to accurately monitor the changes in concentration (Treado et al, 2006).

3.1.1.1 Mercuric Chloride Analysis in Pipe Sediment

An automated CVAAS, the Hydra C, was used to measure mercury content in both experimental solutions and pipe materials. In this mercury (Hg) analyzer the sample is combusted at high temperatures with oxygen. Gases are carried through a heated catalyst tube that removes halogens, nitrogen oxides, and sulfur oxides, and the remaining products are carried through a gold amalgamation tube. All Hg is captured, heated, and released as a gaseous bolus toward the CVAAS. The signal is measured in series by a high sensitivity cell followed by a low sensitivity cell, and the two peaks are integrated and reported against a calibration.

This instrument was the manufacturer's first generation and several issues were revealed upon method development. Samples are analyzed in sample boats which consisted of nickel metal. The nickel metal was highly susceptible to corrosion from nitric acid solutions and high temperature used during the analysis. Constant examination of the nickel boats was necessary to remove any damaged boats (holes present). The manufacturer also advertised the availability of ceramic boats, but final production method for such boats had not been perfected at the time of this purchase. A small set of 'trial' ceramic boats were tested and found to be defective: non-even ceramic coating on nickel metal substrate and severe cracks in ceramic coating before and after use in the Hydra C. It was found that small amounts of samples were necessary to avoid saturating the instrument detector. Serial dilution of liquid samples was required and if not carefully performed added much error to the results. Sample carryover also occurred regularly, so two blank sample boats were run in between each experimental sample to prevent such carryover. Solid sample analysis is possible on the Hydra C. A precipitate formed in adsorption experiments using Cu pipe and mercuric chloride caused a poisoning of catalyst and furnace apparatus in the Hydra C. This required replacement of several major parts of the instrument. Sample precipitates could not be analyzed because of these results.

3.1.1.2 *Diffuse Reflectance Infrared Spectroscopy for Substrate Analysis*

Diffuse Reflectance Infrared Fourier Transform Spectroscopy (DRIFTS) analysis provides the vibrational modes of chemically bound constituents much like Fourier-Transform infrared spectroscopy (FT-IR) for fine particles and powders in the concentration range from 0.1 % (by mass) to neat. DRIFTS does not require any special sample preparation; powders can be measured 'as is" or mixed with a diffusely scattering matrix such as potassium bromide. Organic and inorganic materials with a particle size between 2 μm to 5 μm can be analyzed. Solid samples could also be examined using emery paper to remove amounts of the sample of interest; the emery paper can then be analyzed for DRIFTS spectra. DRIFTS measures any changes that occur in the IR beam when the beam interacts with a particulate sample. When the beam reflects off the sample surface, *true specular reflection* is produced and considered as interference in IR detection. When the IR beam penetrates a particle, it scatters or diffuses depending on the angle of the beam; it may also penetrate other particles as it moves through the specimen or reflect off their surfaces. Diffused light that travels through and is partially absorbed by the particles of a specimen contain information about absorption characteristics of the specimen and is called *diffuse reflection*. Specular reflections are separated out by analyzing pure backgrounds of the diluting matrix.

Factors that affect diffuse reflection include: refractive index, particle size, homogeneity and concentration of the specimen. Samples with higher refractive indices and higher concentrations produce more specular reflection and require dilution. Larger particle sizes increase specular reflection; grinding samples to reduce particle size before analysis can eliminate problems. More uniform samples produce more linear relationships between band intensity and sample concentration. So thorough mixing is necessary. A sample holder or cup is often used in DRIFTS accessories. To properly prepare a powder sample for maximum signal, the powder should overflow the cup. The powder is then leveled with a spatula to reduce reflection from the sample surface. The sample should not be tamped as particles with too close proximity reduce IR beam penetration. The rule for diluting (by weight) samples is: for organic samples- 10 % of sample is mixed with 90 % of diluting matrix and for inorganic samples- 2 % to 5 % of sample is mixed with 95 % to 98 % of diluting matrix.

Since water produces broad peaks at high wave numbers in a FT-IR spectrum and can obscure many peaks of interest produced from chemical contaminants, all powder samples from adsorption experiments has to be filtered and dried. Experiments using physical additions of organic contaminants to $CaCO_3$ showed new peaks that were characteristic of the contaminant. Powder $CaCO_3$ samples used in adsorption experiments were filtered using a house vacuum. This procedure appeared to remove much of the volatile organic contaminants as no discernable signal was observed. In addition, DRIFTS did not provide any useful information for the cyanide experiment on

Cu pipe, precipitates that formed, or $CaCO_3$. For this study, DRIFTS could not easily characterize the hard copper pipe substrates.

3.1.1.3 Infrared Micro Spectroscopy for Substrate Analysis

Infrared (IR) micro spectroscopy in reflectance mode was used to examine Cu pipe materials before and after phorate adsorption experiments by and after each decontamination step to determine the extent of phorate removal. The resulting reflection-transmission spectra are equivalent to traditional absorption spectra after converting them to absorbance without any mathematical correction. The IR micro spectroscopy was performed with a Thermo Scientific/Nicolet Magna-IR550 FTIR spectrophotometer interfaced with a Nic-Plan IR microscope. The microscope was equipped with a video camera, a liquid nitrogen-cooled mercury cadmium telluride detector (Thermo Scientific, Inc.; Madison, WI), and a computer-controlled mapping translation stage (Spectra-Tech, Inc.; Shelton, CT), which is programmable in the x and y directions. IR micro spectroscopy testing consisted of individual FTIR measurements of 3 to 5 spots on the pipe specimen and representative FTIR mapping if obtainable. The individual spectra were taken from the middle and sides of the specimens. Usually, 5 spectra were taken if the 1st 3 spectra differed from each other. For most experiments, representative, equally spaced spots on the Cu pipe samples were analyzed in the IR spectroscopy mode. If the contaminant layer and pipe sample were sufficient for IR mapping, an IR map was collected either based on a portion of the contaminant IR spectrum or on a particular IR peak that represented the chemical contaminant. Spectral point-by-point mapping of the pipe materials was done in a grid pattern with a computer controlled microscope stage and Atlus software package. Spectra were collected from 4000 cm^{-1} to 650 cm^{-1} at the spectral resolution of 8 cm^{-1} with 32 scans and a beam spot size of 400 μm x 400 μm. The spectra were normalized to the background of bare spots on the pipe material and were successively measured during the mapping after ever 20 spectra to compensate for slight changes. The resulting images were displayed as color contour maps in the desired region.

For all of the IR micro spectroscopy measurements, all spectra were collected at the *same exact spot* on the pipe samples after the various treatments. Different morphologies (more or less deposits on the Cu substrate) on the pipe samples were also purposefully chosen for IR analysis. The IR spectrum for pure phorate was also analyzed for comparison. IR spectra of typical Cu in-service pipe were collected and showed that all the IR spectra were different for the several spots on the pipe. This was an indication of the heterogeneity of the deposits on the surface of the pipe. Furthermore, broadness of the IR peaks indicated the complex composition of the deposits on the pipe sample at several spots. After phorate adsorption onto the Cu in-service pipe, peaks that were characteristic of phorate appear as IR bands at approximately 1375 cm^{-1}, 1100 cm^{-1}, 1005 cm^{-1}, 950 cm^{-1}, and 795 cm^{-1} as shown in the IR spectrum. IR spectra were then examined for significant differences between phorate exposed pipe samples after water decontamination. PVC pipe samples could not be successfully examined by IR micro spectroscopy because the PVC material caused interfering IR peaks that obscured any contaminant peaks.

3.1.1.4 Raman Microscopy for Pipe Substrate Analysis

Raman spectroscopy is used to determine molecular structures and compositions of organic and inorganic materials in a similar manner to IR spectroscopy. When an intense beam of monochromatic light from a laser impinges on a material, scattering can occur in all directions with the frequency of the scattered light the same as the original light; this effect is known as *Rayleigh scattering*. Another type of scattering that can occur simultaneously with Rayleigh scattering is known as the *Raman effect*. It occurs at frequencies both higher and lower than the original light with considerably diminished intensities. The difference between the incident and scattered frequencies are equal to the actual vibrational frequencies of the material and are characteristic to the chemical functional groups in a material. Raman is particularly useful in examining aqueous solutions of inorganic and organic compounds. Water is a poor Raman scatterer, so observations of vibrational transitions normally obscured in intense water adsorptions in IR spectroscopy are allowed.

For the adsorption and decontamination experiments, a Raman microscope (Bruker Senterra) was used to collect Raman spectra on 3 to 5 spots of Cu or PVC pipe. The locations were spaced out to achieve the best representation of the sample. Analysis began with a 50x objective (longer working distance) and then to a 100x objective. The location of the x, y stage was recorded for each analysis and optical images of the locations were also saved. The basic analysis protocol was to determine if the spectra look the same for all locations on clean pipe materials (rinsed in methanol) and to define the background for the pipe material. For original clean pipe, all spectra were similar.

The same pipe sample was then subjected to a drop of fuel or other contaminant and analyzed using the same protocol to determine the presence of new peaks from the fuel. Both fuel evaporation in air and the power of the laser affected the stability of the new Raman signals. This procedure was used to determine instrument parameters (type of laser light, laser power, resolution [range of wave numbers to be examined], and laser dwell time) for each pipe sample material and contaminant. The choice of laser power was based on obtaining good Raman signal (low signal to noise) with no sample damage. High laser power burned samples. For reflective samples lower laser power was required to avoid saturating the detector. A lower dwell time and shorter range of wave numbers decreased the overall analysis time, which reduced the probability of contaminant evaporation.

These instrument parameters were then used as a starting point on pipe materials subjected to adsorption and decontamination experiments. In general, Raman was used to make qualitative identification of the chemical contaminant. For in-service Cu pipe, very low laser power was necessary to avoid burning the original deposits; therefore, the resulting Raman signals were very weak. It was difficult to achieve quantitative information due to the evaporation of the contaminant in air and by the instrument laser. The data collected from Raman analysis provided optical images of the pipe, 3-D maps of Raman spectra (particular peaks related to specific contaminants), and individual spectra from the pipe materials.

3.1.1.5 X-ray Photoelectron Spectroscopy and Scanning Electron Microscopy

After the adsorption experiments for residual contaminant, the pipe and powder substrates were examined using X-ray photoelectron spectroscopy (XPS). XPS is a very surface sensitive analytical technique that examines the top 5 nm of the surface and has a detection limit of about 1 % atomic concentration. It provides quantitative elemental and chemical species information. All XPS measurements were performed with a Kratos Axis Ultra photoelectron spectrometer. Experiments were conducted at room temperature with a base pressure in the $1.3x\ 10^{-6}$ Pa range. The monochromatic Al Kα x-ray source was operated at 140 W (14 kV, 10 mA). The energy scale was calibrated with reference to the Cu $2p_{3/2}$ and Ag $3d_{5/2}$ peaks at binding energies (BE) of 932.7 eV and 368.3 eV. A coaxial charge neutralization system provided charge compensation. Pipe samples and pipe precipitates were attached to the XPS sample holder using double stick tape. The analysis area for the high-resolution spectra was 2 mm x 1 mm.

A survey scan was performed on all samples to determine the elements of interest, In general, O 1s, N 1s, C 1s, S 2p, P 2p, Ca 2p and Cu 2p spectra were acquired at a pass energy (PE) of 20 eV and a maximum acquisition time of 8 minutes per element. Peak BEs were determined by referencing to the adventitious C 1s photoelectron peak at 285.0 eV. Quantitative XPS analysis was performed with the Kratos VISION software (version 2.1.2). The atomic concentrations were calculated from the photoelectron peak areas by subtracting a linear-type background. The O 1s, N 1s, and C 1s regions were deconvoluted using mixed 70 % Gaussian/30 % Lorentzian components.

Challenges with XPS are the high vacuum requirements which limit all samples to dry conditions. In addition, volatile contaminants that were not chemically reacted to the surface also have a greater possibility of being pumped away before analysis. In general, increases in carbon, phosphorus, and sulfur were observed. Due to controlled access of the instrument and high vacuum requirements, XPS was not the best choice for most surface analysis.

Scanning electron microscopy (SEM) was also used to examine pipe materials before and after adsorption experiments. SEM is primarily used to study the surface topography of solid samples and has a resolution of 1.5 nm to 3.0 nm. Electrically conductive materials may be examined directly, but non conductive materials require a thin conductive coating (carbon and precious metals) to prevent electrical charging of the specimen. An electron beam passing through an evacuated column is focused by electro-magnetic lenses onto the specimen surface. The beam is then rastered over the specimen in synchrony with the beam of a cathode ray display screen. The secondary electron emission (inelastically scattered) from the sample is then used to modulate the brightness of the cathode ray display screen, thereby forming the image. If back-scattered electrons (elastically scattered) are used to form the image, the image contrast is determined largely by compositional differences in the sample surface rather than topographic features. In this study, the pipe material had to be dried for SEM analysis. Depending on the material, a carbon coating was used for non-conductive samples. Secondary electrons were generally used to collect images and energy dispersive X-ray analysis (EDS or EDX) was used for element identification.

3.1.2 Biological Contaminants

3.1.2.1 Biofilm Reactor Measurements

During the course of the project, three different configurations were used to establish biofilms on plumbing materials. The first configuration used was the CDC (Centers for Disease Control) biofilm reactors (obtained from BioSurface Technologies, Bozeman, MT). The second configuration was a pipe section reactor constructed in a lab that was operated as a single flow through with creeping flow (1 mL/min). The third configuration was the pipe section reactor operated in a loop mode with intermittent high flow rate (2.5 L/min, 2 h with flow and 2 h without flow in a continuous cycle). The first two configurations used synthetic water supplemented with humic acids. The third used local (Gaithersburg, MD) tap water supplemented with humic acids to establish the biofilm layers.

The CDC biofilm reactor has 24 coupon disks (13 mm diameter) made of PVC or copper (coupons) suspended in a 1 L beaker. A Teflon baffle is suspended in the reactor and the stirring rate of the baffle results in the fluid shear on the coupon surfaces. A separate peristaltic pump was used to add the synthetic water solution to the CDC reactor. The CDC reactor was operated without stirring for the first week and for the second and third week the baffle was maintained at 120 rpm (Morrow et al., 2008).

The pipe section reactor used a series of plumbing pipes (PVC and copper) with 19 mm (¾ inch)^2 diameter and 51 mm in length. The pipe sections were jointed together using silicone tubing. Peristaltic pumps were used to control the flow of water in the pipe section reactors. The third configuration used to grow biofilms on pipe surfaces was the pipe section reactor that was in a loop and local tap water supplemented with humic acids was used to establish the biofilms. This third configuration used an intermittent high flow (2 h on and 2 h off) to better simulate the type of conditions found in a building water system.

3.1.2.2 Inactivation of Biofilm Spores on Pipe Materials

A Teflon baffle was used to establish the fluid shear in the CDC biofilm reactor. Peristaltic pumps were used to control the fluid velocity in the pipe section reactor (second and third configurations). When using the CDC biofilm reactor, the biofilm conditioned coupons were contacted with spores by adding the spores to the reactor under two shear conditions (one without baffle stirring and one with the baffle stirring at 60 rpm). After contacting with spores, the coupons were then removed from the reactor, rinsed, placed in a tube, and then contacted with the disinfectant solutions without stirring.

---

[2] Use of Non-SI Units in a NIST Publication: The policy of the National Institute of Standards and Technology is to use the International System of Units (metric units) in all of its publications. However, in North America in the heating, ventilation and air-conditioning industry, certain non-SI units are so widely used instead of SI units that it is more practical and less confusing to include some measurement values in customary units only.

The spore solutions with a concentration of approximately 10^7 CFU/mL were contacted with the biofilm conditioned pipe coupons or pipe sections. The contact time varied from 2 h to 24 h. The adhesion of the spores reached a plateau at approximately 24 h. After contacting the spores, the coupons and pipe sections were rinsed with water. The concentration of the spores adhered to the coupon or pipe section were determined by sampling several coupons or pipe sections at this stage. The coupons and pipe sections with spores adhered to the biofilm were then contacted with disinfectant solutions. Different concentrations of active chlorine or monochloramine were contacted with the coupons or pipe sections varying period of time. The coupons or pipe section were removed from the disinfectant solution, rinsed with 7.5 mM thiosulfate to stop the reaction of chlorine or monochloramine, and then rinsed in water before sampling.

In the second configuration the pipe section reactor was operated with a creeping flow (1 mL/min) for biofilm growth (three weeks). After biofilm growth stage, the pipe sections were disassembled and the individual pipe sections were contacted with spore solutions, rinsed and then contacted with disinfection solutions. The contacting and disinfection steps were done without flow.

A major difference in the conditions of spore contacting, flushing, and disinfection was used in the third configuration compared to the first two configurations. In the third configuration, the pipe sections operated in a loop mode with intermittent high flow for the spore solution (2×10^6 CFU/mL in 2 L tap water) contacting phase for 24 h (cycling 2 h on and 2 h off). The water flushing and disinfection with chlorine solutions stages was done at high flow (2.5 L/ h with continuous flow).

3.1.2.3 Inactivation of Ricin by Disinfectants

Ricin has native fluorescence due to the presence of aromatic amino acids in the subunits (Gaigalas et al., 2007). Trytophan has the highest fluorescence of the amino acids. Chlorine and monochloramine are chemical oxidants known to react with many amino acids, including trytophan (Nightingale et al., 2000; Hawkins et al., 2003; Alimova et al., 2005). The native fluorescence of ricin was monitored by exciting the samples with light at 280 nm and measuring the fluorescence emission light at 340 nm. Fluorescence is easily monitored property that can be done in real time and in a noninvasive or destructive manner. The biological activity of ricin using cell culture (described above) was measured. The biological activity of ricin and the fluorescence of the same samples were also measured to determine if the loss of fluorescence was correlated to the loss of biological activity.

Since ricin is an enzyme, the inactivation of two model enzymes, lactate dehydrogenase (LDH, from cow heart) and lysozyme (from chicken egg) were studied. The native fluorescence of these two model proteins was monitored along with their enzymatic activity. Cow heart LDH and chicken egg lysozyme were chosen because these proteins can be obtained in a pure form for reasonable costs and the enzymatic activity can rapidly be measured.

3.2 Dynamic Fluid/Surface Interface Measurements

A fluorescence based measurement technique was developed to measure diesel fuel adsorption to polyvinyl chloride (PVC), iron, and copper during contaminated water flow and tap water flushing (Kedzierski, 2006 and Kedzierski, 2008). The test apparatus was designed for the purpose of studying adsorption of diesel fuel from a flowing dilute diesel/water mixture. It was used to measure the mass of diesel fuel adsorbed per unit surface area (the excess surface density) and the bulk concentration of the diesel fuel in the flow. Both bulk composition and the excess surface density measurements were achieved via a traverse of the fluorescent measurement probe perpendicular to the test surface. The diesel adsorption to each test surface was examined for three different Reynolds numbers[3] between zero and 7000. Measurements for a given condition were made over a period of approximately 200 h for a diesel mass fraction of approximately 0.15 % in tap water. The adsorbed diesel on the surfaces was removed by flushing with tap water. Excess surface density measurements were made during flushing.

3.2.1 Experimental Apparatus and Measurement Uncertainties

A test apparatus was designed and developed to use the fluorescent properties of diesel fuel to study its adsorption and desorption to and from plumbing pipe materials. A calibration technique was developed to measure both the mass of diesel adsorbed per unit surface area (the excess surface density) and the bulk concentration of the diesel fuel in the flow. The flow loop for measuring diesel fuel on pipe substrates and the development of the measurement technique with its uncertainties is thoroughly discussed in Kedzierski (2006, 2008).

Figure 3.2.1.1 schematically shows the flow loop for measuring diesel fuel on pipe substrates. The primary components of the loop are the pump, the reservoir, and the test chamber with the test section. The inside surfaces of the approximately 96 mm x 1.6 mm rectangular flow cross-section of the aluminum test chamber, shown in Fig. 3.2.1.2, were black anodized to minimize stray light reflections. A centrifugal pump delivered the contaminated water to the entrance of the rectangular test chamber at room temperature. The flow rate was controlled and varied by varying the pump speed with a frequency inverter. A heat exchanger immersed in the reservoir was supplied with brine from a temperature-controlled bath to maintain the entrance temperature to the test chamber at ambient temperature (293.8 K). This was done to ensure that the diesel fuel was at the same temperature as it was during the fluorescence calibration to avoid the temperature effect on fluorescence (Miller, 1981). An additional temperature-controlled bath was used to maintain the fluorescence standards at the same ambient temperature. As described in this section, the fluorescence standards were used to calibrate the range of the fluorescence measurements.

Residential copper pipe was used to plumb together the various components of the loop. Redundant volume flow rate measurements were made with an ultrasonic doppler and a

---

[3] The Reynolds number is a dimensionless number that characterizes the swiftness and randomness of the flow. Larger Reynolds numbers indicate more random and swifter flow.

turbine flowmeter with expanded uncertainties of $\pm$ 0.12 m^3/h and $\pm$ 0.03 m^3/h, respectively. As shown in Fig. 3.2.1.1, three water pressure taps before and after the test chamber permitted the measurement of the upstream absolute pressure and the pressure drops along the test section with expanded uncertainties of $\pm$ 0.24 kPa and $\pm$ 1.5 kPa, respectively. Also, a sheathed thermocouple measured the water temperature at each end of the test chamber to within an uncertainty of $\pm$ 0.25 K. The dissolved oxygen level, the conductivity, and the pH, were monitored at the water reservoir with associated B-type uncertainties of $\pm$ 0.5 %, $\pm$ 50 $\mu\Omega$/cm, and $\pm$ 0.3, respectively.

Figure 3.2.1.1 also shows the inlet and exit taps used to flush the test section with fresh tap water. In preparation for flushing, the test section was isolated from the rest of the test loop by closing valves. Then the fluid was drained from the test chamber and returned to the reservoir. Next, a tap water supply was connected to a test chamber port. The other test chamber port was connected to a filter to absorb any diesel fuel before it was sent to a drain.

Figure 3.2.1.2 shows a view of the spectrofluorometer that was used to make the fluorescence measurements and the test chamber with the fluorescence probe perpendicular to the flattened pipe test surface. The spectrofluorometer was modified by replacing the cuvette holder with a special adapter with lenses and mirrors to remotely excite and measure fluorescence via a bifurcated optical bundle. Two optical bundles consisting of 84 fibers each originated from the spectrofluorometer. One of the bundles transmitted the excitation light, i.e., the incident intensity (I_o), to the test pipe surface. The other bundle carried the emission, i.e., the fluorescence intensity (F), from the test surface to the spectrofluorometer. The excitation wavelength (λ_x) and the emission/detection wavelength (λ_m) were set to 434 nm and 485 nm, respectively, for all tests. Further details on the fluorescence measurement technique are given in Kedzierski (2006, 2008).

Number 2 diesel fuel was used from a single batch throughout the experiment to avoid property variations that might be caused by batch variations due to it being a complex mixture of hydrocarbons. A nominally 1.5 % by mass diesel mixture was prepared with local Gaithersburg, MD tap water for the exposure/flow rate tests. The measured dissolved oxygen level, the conductivity, and the pH, of the water at 24 °C before mixing with diesel fuel were found to be, 86.4 %, 358 $\mu\Omega$/cm, and 7.04, respectively.

During contamination tests, diesel adsorbs to the test surface to an excess layer thickness (l_e). Because the molar mass of the diesel is unknown, the surface excess density (Γ) is defined in this work on a mass basis as (Kedzierski, 2002):

$$\Gamma = l_e (\rho_d - \rho_b x_b) \qquad (3.2.1.1)$$

The density of liquid diesel fuel is ρ_d. The density of the flowing bulk mixture (ρ_b) is evaluated at the bulk mass fraction of the mixture (x_b). The surface excess density is roughly the mass of diesel attached per surface area. The Γ and l_e are the primary measurements of this study. The l_e is measured perpendicular to the surface with the origin at the fluid-surface interface.

Two different calibration methods had to be combined due to the additional complexity caused by immiscible liquids. Both calibration techniques were used to quantify different functional aspects of the Beer-Lambert-Bougher law (Amadeo et al., 1971), which forms the basis of the calibration equation. The first method was used to obtain the relationship between diesel composition and fluorescence intensity for a fixed light path length (fixed probe height above the test surface). The first method would have been sufficient had the bulk composition of the flow remained the same as it was charged in the reservoir. Due to the immiscibility of the two fluids, the bulk composition of the flow differs from that in the reservoir. As a result, a second method is necessary to determine both the contaminant mass fraction and the excess layer thickness. The second method that was developed in this study relies on a perpendicular traverse of the flow stream with the measurement probe. To achieve this, a linear positioning device with a graduated knob was adapted to the quartz tube as shown in Fig. 3.2.1.2. The second method (traverse method) is used to calibrate the effect of contaminant thickness (path length) and the proximity of incident intensity. The traverse method is essential for splitting the total measured fluorescent intensity into two components: that from the diesel fuel on the test surface and that from the diesel in the bulk flow stream. In this way, the mass the diesel fuel on the test surface and the composition of the fluid stream are determined.

Two standard jars were used as reference standards to set the lower (0) and upper (100) limits of the intensity signal on the spectrofluorometer for raw measurements made at the test section (F_r). A jar that contained only pure water was used to zero the intensity. A second jar that contained pure diesel fuel was used to set the intensity on the spectrofluorometer to 100. All raw-measured intensities (F_r) were numerically normalized by the intensity from the zero-jar (F_0) and the maximum-jar (F_{100}) as:

$$F = \frac{F_r - F_0}{F_{100} - F_0}$$
(3.2.1.2)

where the intensity of the contamination data was adjusted by no more than 0.3 % to account for the small (typically within ± 1 K) difference in temperature between the test section and the bath containing the maximum- and the zero-jars (Kedzierski, 2006). The maximum correction for the flushing data was approximately 1.5 %, which was larger than for the contamination measurements due to the colder temperature of the house tap water.

The linear form of the Beer-Lambert-Bougher law (Amadeo et al., 1971) shows that the measured fluorescence intensity is related to the incident light intensity (I_o), the extinction coefficient (ε), the concentration of the fluorescent diesel (c), the path length of light (l), and the quantum efficiency of the fluorescence (Φ) as:

$$F = 2.3 I_o \varepsilon c l \Phi \rightarrow [\varepsilon c l \leq 0.05]$$
(3.2.1.3)

27

The linear criteria for eq. (3) ($\varepsilon cl \leq 0.05$) is satisfied for 78 % of the calibration data, and the absorbance (εcl) did not exceed 0.063 for all of the data. As a result, the calibration measurements gave: $2.3 I_o \Phi \varepsilon M_c^{-1} = 1.04735 \left[\text{m}^2 \text{kg}^{-1} \right] e^{-209.23[\text{m}^{-1}]l}$ when fitted to eq. (3.2.1.3) (Kedzierski, 2006). Using the calibration and expressing the concentration in terms of the bulk mass fraction and the bulk liquid density gives the calibration of the fluorescence intensity in terms of the mass fraction and path length as:

$$F = \frac{2.3 I_o \Phi \varepsilon}{M_c} l x_b \rho_b = 1.04735 \left[\frac{\text{m}^2}{\text{kg}} \right] l x_b \rho_b e^{-209.23[\text{m}^{-1}]l} \qquad (3.2.1.4)$$

Note that the concentration of the fluorescent diesel fuel has been replaced with the product of the bulk contaminant (diesel) mass fraction (x_b) and the density of the bulk mixture (ρ_b) divided by the molar mass of the contaminant (M_c). The mixture densities were calculated from a linear mass weighted basis of the pure fluid specific volumes.

Because the actual concentration of the diesel fuel entrained in the water flow stream is unknown, eq. (3.2.1.4) cannot be directly used to obtain the excess layer of diesel on the pipe surface. The Γ and the l_e must be obtained from additional information that is obtained from a perpendicular traverse of the flow stream. As shown in Fig. 3.2.1.2, a linear positioning device with a graduated knob was used to traverse and locate the quartz tube relative to the test surface and thus measure the path length of the incident light through the fluid. Measurements of the fluorescence intensity (F) for various path lengths provided sufficient information for obtaining both the bulk mass fraction and the excess layer thickness. The methodology for this is explained in the following.

The total fluorescence signal (F) can be separated into three components along the path length while assuming a uniform bulk mass fraction. The total intensity is the sum of that contributed by the bulk concentration ($F_l(x_m = x_b)$) for the entire path length and that in the diesel excess layer ($F_{le}(x_m = 1)$) minus the intensity that would have been due to the bulk concentration but did not occur because it was displaced by the excess layer ($F_{le}(x_m = x_b)$)

$$F = F_l(x_m = x_b) - F_{l_e}(x_m = x_b) + F_{l_e}(x_m = 1) \qquad (3.2.1.5)$$

Substitution of eq. (3.2.1.4) into the components of the above equation and grouping like terms gives:

$$F = 2.3 I_o \Phi \varepsilon M_c^{-1} \left[l x_b \rho_b - l_e x_b \rho_b + l_e \rho_d \right] \qquad (3.2.1.6)$$

Here ρ_d is the density of liquid diesel.

For a given probe traverse, the only variable in eq. (3.2.1.6) is the path length. Consequently, eq. (3.2.1.6) can be arranged in terms of two regression constants for a single traverse:

$$F = \left(A_0 + A_1 l \right) e^{-209.23[\mathrm{m}^{-1}]l} \tag{3.2.1.7}$$

Comparison of eqs. (3.2.1.6) and (3.2.1.7), yields the bulk mass fraction as:

$$x_b = \frac{A_1 e^{-209.23\,\mathrm{m}^{-1}l}}{\rho_d 2.3 I_o \Phi \varepsilon M_c^{-1}} = \frac{A_1}{1.04735 \left[\mathrm{m}^2 \mathrm{kg}^{-1} \right] \rho_d} \tag{3.2.1.8}$$

and the excess layer thickness as:

$$l_e = \frac{A_0}{\rho_d e^{209.23\,\mathrm{m}^{-1}l} 2.3 I_o \Phi \varepsilon M_c^{-1} - A_1} = \frac{A_0}{1.04735 \left[\mathrm{m}^2 \mathrm{kg}^{-1} \right] \rho_d - A_1} \tag{3.2.1.9[4]}$$

The average uncertainty of l_e for the measurements with the iron, the PVC, and the copper surfaces for the contamination measurements was approximately ± 0.06 μm, ± 0.1 μm, and ± 0.2 μm, respectively. The average uncertainty of x_b was approximately ± 0.00008.

The diesel bulk mass fraction of the tap water used during the flushing tests is zero. For flushing tests, eq. (3.2.1.8) produced a non-zero bulk mass fraction with a magnitude close to the uncertainty of the measurement, i.e., typically 0.008 % (80 ppm). An alternative approach for the flushing tests that forces the bulk mass fraction to be zero is to start by setting $x_b = 0$ in eq. (3.2.1.6) and taking its derivative with respect to the path length.

Rearranging the resulting differentiated equation and solving for the excess layer thickness yields:

$$l_e = \frac{\dfrac{dF}{dl}}{209.23[\mathrm{m}^{-1}]\rho_d 2.3 I_o \Phi \varepsilon M_c^{-1}} = \frac{A_0 - A_1 \left(\dfrac{1}{209.23[\mathrm{m}^{-1}]} - l \right)}{1.04735 \left[\mathrm{m}^2 \mathrm{kg}^{-1} \right] \rho_d} \tag{3.2.1.10}$$

Equations (3.2.1.9) and (3.2.1.10) are equivalent for negligible A_1, which is the case for flushing. However, eq. (3.2.1.10) was used to obtain the l_e for all of the flushing measurements because of its more explicit derivation. The average value of l_e was used

---

[4] The methodology presented here is a refinement of that given in Kedzierski (2006). Equations (3.2.1.8) and (3.2.1.9) are more explicit than those presented in Kedzierski (2006), but give the same results.

for a given measurement probe traverse. The average uncertainty in l_e for the iron and the PVC flushing tests was approximately ± 0.02 μm, and ± 0.05 μm, respectively.

3.3 Screening Tests

The purpose of the screening tests was three fold. First, due to the large number of contaminant/substrate combinations and the time required to conduct each full scale test, it was deemed desirable to focus on the combinations that would be most useful in providing information to support recommendations. In other words, rather than trying to exhaustively test every combination, which would not have been possible in any case given the time constraints, the screening tests would identify the contaminant/substrate combinations that showed the potential for accumulation, allowing those to be tested in the full-scale test systems. The second purpose for the screening tests was to identify contaminant/substrate combinations that might result in damage to the full-scale plumbing loop. This could result in the need for time-consuming and costly repairs or cause other troublesome measurement problems, such as odor or disposal issues. The third purpose for the screening tests was to provide a broader range of measurement results to help generalize the decontamination recommendations, albeit with data from smaller scale measurements.

High concentrations (mass fractions) of contaminants) were placed in small jars, and a small piece (coupon) of each of the test materials was separately placed in a jar and sealed. See details in Table 3.3.1. Following a certain period of time which varied based on scheduling constraints, the test material was removed from each jar and evaluated for the presence of contaminant. This was followed by flushing with clean tap water for one hour and a recheck for contaminant, and then another clean tap water flush for 24 h and a final measurement. The flushing methodology involved inserting the exposed coupon into a piece of flexible hose through which cold tap water was directed at a flow rate of approximately 0.25 L/s (4 gpm), corresponding to a Reynolds number of approximately 30,000.

Table 3.3.1		*Contaminant Mass Fractions for Screening Tests*			
Contaminant	**Diesel**	**Gasoline**	**Strychnine**	**Cyanide**	**Phorate**
Concentration	100 %	100 %	0.5 %	1 %	100 %

These tests represent a somewhat severe exposure scenario, since the contaminant remains in contact with the substrate for an extended period of time, but one might occur near the point of contaminant introduction in an actual event. Because the objective was to ensure the removal of any significant amount of accumulated contaminants, this worst-case analysis is appropriate.

The presence of contaminant on the exposed coupons was evaluated using a Raman spectrometer to look for the characteristic Raman spectral signatures associated with each of the contaminants. These signatures were determined by first obtaining Raman spectra of each contaminant, along with Raman spectra of each unexposed substrate material. Then, the Raman spectra of the exposed substrates were compared to the baseline and contaminant spectra, and the presence of the contaminant inferred by the observation of characteristic Raman peaks, and the magnitude of any contaminant accumulation estimated by measuring the height of specific characteristic peaks.

Lacking specific calibration data to relate Raman intensity, represented by peak height, to mass or volume of contaminant, consider the contaminant accumulation values to be relative values. However, a later section will attempt to estimate the thickness of the layer of accumulated contaminant for some of the measurements. Note that acceptable levels of residual contaminants within water supply systems is a public health matter that is beyond the scope of this analysis. However, estimation of residual contamination and the methods for effectively removing contaminants is the focus of this work.

3.4 Full-Scale Dynamic Tests

Tests were conducted at the BFRL Plumbing Test Facility, which emulates a full-scale building water supply system in a controlled laboratory setting. The test facility includes water supply piping, fittings, and fixtures that represent a five-story building. It also incorporates a computer-based control and data acquisition system which can be programmed to circulate water according to any desired profile. The facility includes a measurement station for water tanks, such as hot water heaters. Figure 3.4.1 presents a schematic diagram of the full-scale plumbing system test facility.

The types of tests that were conducted involved introducing a water/contaminant mixture into the water distribution system, and allowing it to circulate and/or stand for a set period of time, followed by a flushing or cleaning operation. Pipe and water samples were collected at various stages of each test run and then analyzed for the presence of accumulated contaminant. In this manner, the tendency of different contaminants to accumulate on exposed components of building plumbing systems could be determined along with the effectiveness of flushing and cleaning procedures.

3.4.1 Building Plumbing System Loops

The BFRL Plumbing Test Facility was used to measure contaminant accumulation and removal in full-scale building plumbing systems, including pipe loops and hot water heater tanks. This facility has been described in detail previously in Treado et al. (2006) and Treado (2007). The testing methodology is briefly summarized as follows:

1. Water/contaminant mixtures were prepared in a large tank, and then circulated by pumping through a simulated three floor building plumbing system, which included measurement sections with small removable pieces of copper (3/4 and ½ inch) and PVC (1/2 inch) pipe. (Exception: For cyanide and strychnine, the pipe samples were exposed to the contaminant by suspension in a small container, and then inserted into the pipe loop for flushing, so that the entire test loop would not be subject to contaminant exposure).

2. Pipe samples were removed after exposure for measurement using a Raman spectrometer.

3. The plumbing system was then flushed with clean, cold tap water, at a flow rate of 0.25 L/s (4 gpm) at the pipe samples, and additional pipe samples removed periodically and measured.

4. At the conclusion of the flushing tests, all components of the pipe loop were thoroughly cleaned as needed with hot water and detergent.

Tests were conducted using diesel fuel, strychnine, sodium cyanide and Bacillus thuringiensis (BT) spores, but the sodium cyanide samples did not provide any useful results. Figure 3.4.2 shows a schematic of the test loop. While the figure indicates only a single sink, in actuality there were three sinks on each of three levels, served by different types and sizes of piping materials.

3.4.2 Hot Water Heaters

Tests were conducted with previously used nominally 189 L (50 gallon) hot water heaters using the following contaminants:

- Diesel fuel;

- Strychnine;

- Sodium Cyanide; and

- Bacillus thuringiensis (BT) Spores.

As with the pipe loop tests, water/contaminant mixtures were made and added to empty hot water heater tanks, then allowed to soak, followed by flushing and/or draining and refilling with clean tap water. Periodic samples were taken of sediments from the bottom of the tanks and tank water. Except for the second diesel fuel test, the tank heating elements were not energized. In some cases, multiple Raman measurements were made on each sample or multiple samples, so data are shown as maximum, minimum and average readings. This occurs because contaminant accumulation is not uniform, and the measurement area of the Raman system is very small. The maximum readings are of greatest significance. Ensuring adequate removal of accumulated contaminants wherever they are located is of interest. Figure 3.4.3 shows a schematic of the hot water heater testing apparatus.

4.0 MEASUREMENT RESULTS
4.1 Bench Scale Tests
4.1.1 Chemical Contaminants
Detailed results of bench scale tests of water mixtures with phorate, toluene, gasoline, diesel fuel, strychnine, cyanide salts, and mercuric chloride are provided in Subsections 4.1.1.1 through 4.1.1.6.

Table 4.1.1 below summarizes the results for static adsorption experiments and subsequent decontamination procedures. These results can be used as a general guide for worst case chemical contamination; however, no realistic pipe-flow conditions were employed making this guide limited. In general, all contaminants physically stick to most of the tested pipe substrates, but pipe composition affected the ability of some contaminants to interact. For organic contaminants, the evidence for interaction was simply in the odor of the pipe substrates after adsorption. Analytical methods were also used to examine the pipe substrates for trace amounts of contaminants. Water decontamination of exposed pipe substrates removed a fraction of most of the chemical contaminants and is defined as "partial" in the table. For these decontamination experiments, trace amounts of the chemical were found analytically in the water. In some instances pipe characterization also revealed the presence of the chemical in question. In the cases of organic compounds, a simple odor was also detected in both the water and on the pipe. Decontamination using a bleach solution was also completed for a subset of chemical contaminants. In these cases, only a fraction of the chemical was removed from the pipe and is indicated as "partial" in the table. Trace amounts of the chemical were found analytically in the water. In some instances pipe characterization also revealed the presence of the chemical in question, but in these cases a simple odor was also detected in both the water and on the pipe.

Table 4.1.1 *Summary of Chemical Contaminant Interaction with Pipe Materials and Decontamination Results.*			
Contaminant	**Sticks**	**Removed by Water Flush**	**Removed by Other Additive (specified)**
Diesel fuel	Yes, All materials*	Yes, partial	Bleach solution, partial
Gasoline	Yes, All materials*	Yes, partial	Bleach solution, partial
Toluene	Yes (CuF, CuB, CuIS, PVC, PVCB)	Yes, partial	
Phorate	Yes, All materials*	Yes, partial	Bleach solution, partial
Strychnine	Yes, All materials*	Yes, partial	
Cyanide Salts	Yes (Reacts with all metals; CuF, CuB, CuIS, BR, Fe)	No, reaction product formed on metals	
Mercuric Chloride	Yes, All materials* (reacts with all metals to form Hg metal and can absorb into plastics)	Yes, partial (not advised for metals as Hg metal is present)	

*Pipe materials represent Copper [clean flat (CuF), biofilm growth (CuB), used in hot water heater (CuIS)], PVC [clean (PVC), biofilm growth (PVCB)], iron (Fe), brass (BR), and rubber (RB).

4.1.1.1 Phorate

Phorate adsorption experiments showed an interaction of phorate with all pipe materials [Cu, PVC, calcium carbonate ($CaCO_3$), brass, iron, rubber]. A simple test of odor detection further indicated the presence of phorate. It was assumed that phorate, an organic compound, would absorb more on plastic and rubber pipes than with metal pipes. Phorate in water was analyzed with PT-GC/MS. Concentration profiles from water analysis showed a general decreasing trend in phorate concentration from initial phorate concentration with exposure time to some equilibrium phorate concentration for all pipe materials. This equilibrium phorate value depended on the initial phorate concentration and type of pipe material. GC/MS results showed similar phorate species for control solutions and all pipe materials with no evidence of secondary products. The extent of phorate interaction was determined by comparing the change in concentration between the initial and the equilibrium values. A ranking of decreasing interaction was $CaCO_3$, PVC, Cu pipe with deposits, clean flat Cu. Explanations for the ranking included high surface area of $CaCO_3$, polymeric material interaction with the organic compound, and composition and roughness of the Cu deposit.

Water characterization also indicated an extent of contaminant and pipe material interaction in the water. An increase in TOC concentration was observed, but it never reached initial phorate concentrations, indicating the lack of soluble organic material in water. There were no clear trends in TOC values with pipe material, but phorate/PVC pipe solutions showed the greatest increase in TOC value, an indication of phorate interaction with the PVC pipe. Turbidity values fluctuated over phorate concentration ranges, but an increase was observed with Cu in-service pipes in which visible pipe deposits were observed in the water solution.

Pipe material characterization was essential in determining the fate of phorate contamination on pipes because of reduced solubility of phorate in water. By analyzing the pipe material, deposition of the contaminant can be measured directly to establish a correlation with data from water analysis. Fourier Transform infrared (FTIR) micro spectroscopy data showed that a film of contaminant was present on all Cu pipe samples. IR mapping of the contaminant on pipe materials showed that the layer of phorate was also heterogeneously spread on the surface of the pipe (Figure 4.1.1.1.1). Decontamination of phorate from the pipe required a bleach solution to oxidize the phorate compound or some organic solvent to dissolve phorate from the pipes (Ku, 2002 and Hong, 1998). Decontamination of pipe materials with fresh tap water did remove small amounts of phorate as observed in decontamination solutions using PT-GC/MS for initial phorate concentrations ≥ 24.8 mg/L phorate. FTIR microscopy showed that water rinsing alone did not remove phorate on Cu in-service pipe, but solutions of bleach (as low as 5.3 mL/L) removed the phorate from the surface of the Cu pipe (Fig. 4.1.1.1.1).

4.1.1.2 Toluene

Toluene was chosen as a reference material for gasoline, because it is a component of gasoline. Toluene interacts more strongly with polymeric pipe materials than with metal pipe. PT-GC/MS was used to follow the change in toluene concentration over exposure time for Cu, PVC, and $CaCO_3$). GC/MS analysis showed no evidence of secondary

34

products and species were the same for control solutions and all pipe materials. Toluene concentrations (500 mg/L, 1.9 g/L, 3.8 g/L, 7.5 g/L, 15.1 g/L, and 30.0 g/L) used in these adsorption experiments were much greater than its solubility in water and because it is less dense than water, more toluene may have been lost to the atmosphere upon collecting water samples for analysis. Concentration profiles from PT-GC/MS water analysis showed a general decreasing trend in toluene concentration from the initial toluene concentration with exposure time to some equilibrium value for all pipe materials. This equilibrium toluene value depended on initial toluene concentration and type of pipe material. There were no clear trends between different pipe materials. This may be an indication of the high initial toluene concentration.

Pipe material characterization was used to determine the fate of toluene water mixture on the pipe material. Raman micro spectroscopy was used to examine PVC and Cu pipe materials. Although there was a detectable toluene smell from the pipes, for all Cu pipe materials (clean flat, biofilm growth, and in-service) no Raman signal was detected for toluene. There was no significant difference in the Raman spectra for initial Cu pipes and those exposed to various concentrations of toluene. This could be explained by the volatility of toluene and power intensity of the laser used to obtain Raman signals. Modifying Raman instrument parameters to decrease laser intensity did not change results for Cu pipe samples. PVC pipe material showed visible swelling of the pipe material induced from toluene in water and toluene was observed in the Raman spectrum for 15.1 g/L and 30.0 g/L pipe samples (Fig. 4.1.1.2.1). No toluene was observed on exposed $CaCO_3$ powder using diffuse reflectance Fourier-Transform spectroscopy (DRIFTS). This lack of signal could be due to toluene evaporation from the filtration method used to remove contaminated water for DRIFTS analysis.

Water characterization results were used to follow the extent of contaminant and pipe material interaction in water. TOC concentration did not always increase linearly with increasing toluene concentration and never reached the initial toluene concentration, an indication of the insolubility of toluene in water. PVC exposed pipe materials showed the greatest increase in TOC value. Turbidity values exhibited no clear trend with toluene concentration and no trends were observed in the conductivity, chlorine, and alkalinity methods.

Decontamination of toluene pipes with fresh tap water appeared to remove a fraction of the toluene from the exposed pipe. Analysis of decontaminated water with PT-GC/MS showed that amounts of toluene were present in the decontamination water for all pipe materials. The toluene concentration detected was greatest for PVC pipe samples. The measured decontaminated toluene concentration was about half of the measured initial contamination toluene concentration, regardless of initial toluene contamination concentration. Water characterization of decontamination water showed a slight increase in conductivity and increase in TOC value. Decontaminated pipe material analysis with Raman microscopy did not show any change in the toluene spectrum compared with the initial contaminated spectrum. This is especially important for the PVC samples in which tap water removed a fraction of toluene.

4.1.1.3 Gasoline and Diesel Fuel

Gasoline and diesel fuels were tested separately on Cu (clean flat, with biofilm, and in-service from a hot water heater line), PVC (clean and with biofilm), and $CaCO_3$ pipe materials. Fuel concentrations (500 mg/L, 1000 mg/L, and 2000 mg/L) in these adsorption experiments were much lower than that used in toluene adsorption experiments. PT-GC/MS was used to follow the change in fuel concentration in the fuel water mixture with exposure time to the pipe materials. Each fuel water mixture had a unique gas chromatogram (Fig. 4.1.1.3.1). The chromatogram for fuels consisted of many components, each of which was identified using GC/MS and a MS library search. There was no evidence of secondary products and the species were the same for control solutions and all pipe materials with exposure time. For gasoline experiments, toluene and four components consisting of benzene and benzene derivatives were followed over exposure time. For diesel fuel experiments, benzene and napthalene components were monitored. For both fuels, a slight decrease in peak concentration with exposure time was observed in concentration profiles for all monitored components. Furthermore, there was no significant difference in concentration profiles for control or blank experiments (no pipe materials). These results may indicate that the initial fuel concentrations were too high to measure any adsorption onto the pipe material.

Raman micro spectroscopy was used to determine the fate of fuel water mixture on PVC and Cu pipe materials. Although there was a detectable fuel odor from all exposed pipe materials, for most pipe samples no Raman signal was detected for the fuel. However, the presence of fluorescence was observed for pipe materials exposed to diesel fuel. It should be noted that the biofilm growth on the PVC and Cu pipe materials also caused fluorescence, a potential interference for those pipe materials with biofilm. In general, there was no significant difference in the Raman spectra for the initial Cu pipes and those exposed to the various concentrations of fuel. However, diesel peaks were observed on Cu flat pipe that was exposed to 2000 mg/L diesel (Fig. 4.1.1.3.2). The general lack of Raman signal of the fuel on most pipe samples could be explained by the volatility of the fuel and the power intensity of the laser used to measure Raman signals. Modifying the instrument parameters to decrease laser intensity did not change the results for any of the pipe samples.

Water characterization results were used to follow the extent of contaminant and pipe material interaction in the water. An increase in TOC concentration was observed, but did not increase linearly with increasing fuel concentration and never reached the initial fuel concentration, an indication of the insolubility of the fuel in water. In general, the pH increased with fuel addition, but was not dependent on fuel concentration. A decrease in chlorine content was observed for all adsorption experiments. There was also an increase in alkalinity, but no significant change in conductivity. Turbidity values increased with fuel addition, but no clear trend was observed with pipe material. This increase in turbidity may be due to release of pipe material surface growth or the insolubility of the fuel in water.

Decontamination of fuel exposed Cu and PVC pipes with fresh tap water appeared to remove a fraction of the fuel from the contaminated pipe materials. Bleach solutions were also used to decontaminate both fuels individually from clean flat Cu pipe material. Analysis of the decontaminated tap water with PT-GC/MS showed that amounts of fuel were present in the decontamination water for all pipe materials. The amount of fuel present depended on the time elapsed between contamination (adsorption experiments) and decontamination procedures. All exposed pipe samples were removed from the contaminated water solution and placed in sealed jars without water. For samples immediately decontaminated, more fuel was present in the decontaminated water as shown by PT-GC/MS analysis. The measured concentration of decontaminated fuel ranged from 350 times to 1000 times lower than that measured for the initial contamination fuel concentration. No clear trend was observed for the various initial fuel concentrations and pipe materials because the time elapsed between decontamination varied. PT-GC/MS analysis of decontaminated bleach solutions did not show any peaks for the fuel components.

Water characterization of the decontamination water further confirmed the removal of fuel. TOC values increased with increasing initial contamination fuel concentration, but were on average 7 times lower than the TOC value observed in contamination experiments. The TOC value also depended on the time elapsed between contamination and decontamination, so a greater TOC value was observed for more immediate decontaminations. The pH also increased slightly and the Cl content decreased to near zero. In general, decontaminated pipe material analysis with Raman microscopy did not show any change in the spectra compared to the initial contamination. Only the 2000 mg/L diesel exposed clean flat Cu pipe showed evidence of fuel removal via the disappearance of the diesel Raman peaks and the fluorescence with water decontamination.

4.1.1.4 Strychnine

The entire set of adsorption and decontamination experiments were not carried out for strychnine. Preliminary results of pipe materials analysis using fluorescence imaging was obtained for a "worse case" scenario experiment using 1 g strychnine hydrochloride dihydrate in 65 mL of tap water. Pipe samples consisting of Cu in-service from a hot water heater, brass, iron, rubber and PVC were exposed to the strychnine solution for 2 weeks and were rinsed in 450 mL water before fluorescence imaging analysis (Appendix A). No visible evidence of reaction or interaction was observed (Fig. 4.1.1.4.1a). However, a distinctive fluorescence signal from the strychnine exposed samples was observed. Fluorescence imaging was used to determine percent coverage of the fluorescing signal on the pipe material (Fig. 4.1.1.4.1b) using principal component analysis. Principal component analysis showed that brass and iron were 100 % covered, PVC was 78 % covered, and Cu in-service was 66 % covered. These results show that strychnine, an organic nitrogenous base, does interact or stick to pipe surfaces even after tap water decontamination. Water analysis should be performed to determine the presence of strychnine in the water.

4.1.1.5 Cyanide Salts

Adsorption experiments showed a reaction of CN⁻ with metal pipe materials. This is not surprising as cyanide solutions are used in electroplating applications for brass and zinc.

The reaction was especially strong with Cu pipe, a focus of these adsorption experiments. Metal/ CN⁻ reactions are a potential complication. They may leave CN⁻ and metal CN complexes in the contaminated water, which could produce toxic HCN during decontamination (Ismail, 2009 and Hamid, 2009). The extent of reaction depended on the pipe material and initial CN⁻ concentration. At CN⁻ concentrations of 310 g/L in tap water, a 1g piece of Cu pipe dissolved. Iron and brass pipe samples with similar masses also showed a reaction with this high cyanide concentration and a passivation layer resulted on the surface of the metal, leaving both metal pipes with a silver surface. The rate of change in CN⁻ concentration, as measured by a CN⁻ ion selective electrode (ISE), changed only slightly as a function of surface deposit present on the Cu pipe. For the lowest initial CN⁻ concentration, 3 mg/L, an equilibrium CN⁻ concentration value approached zero. There were also visible differences in the appearance of Cu materials from a shiny orange color to a dull black deposit. The black deposit also precipitated into the water solution. Analytical techniques, X-ray photoelectron spectroscopy (XPS), x-ray diffraction spectrometry (XRD), and scanning electron microscopy (SEM), showed no cyanide species present on the surface of Cu pipe materials. The techniques showed that the black precipitate was copper oxide and a carbonaceous substance.

There was no apparent interaction between CN⁻ and PVC pipe. No apparent reaction occurred with PVC or rubber materials. An adsorption study using 310 g/L CN⁻ resulted in no physical change to the PVC or rubber material, but the CN⁻ solution changed from clear and colorless to clear amber. With CN⁻ concentrations up to 50 mg/L, the CN⁻ ISE response was that of a slight increase in CN⁻ concentration over time, but the signal was very noisy, an indication of the presence of some interfering ion with the CN⁻ ISE probe, particularly Cl⁻.

Water characterization results showed a large increase in conductivity values for all CN⁻ solutions. The conductivity value leveled off to about 16 mS regardless of initial CN⁻ concentration. Typical tap water conductivity was 250 μS. Turbidity for CN⁻ solutions increased at the completion of the experiment. The free and total chlorine did not always approach zero values as was observed for the organic contaminants. The total organic carbon (TOC) concentration increased with increasing CN⁻ concentration, but never approached the initial CN⁻ concentrations. A research group at University of Maryland Baltimore County has developed an ion chromatography method to detect metal/CN complexes in water for this research project. Results showed that metal/CN complexes are present in CN solutions exposed to Cu substrates and the metal/CN complex concentration depended on the initial CN⁻ concentration. Furthermore, the concentration of the metal/CN complex changed over time; it was found that the metal/CN complex decreased with storage age of the experimental solution after the adsorption experiment (Appendix B).

Decontamination by rinsing the exposed Cu and PVC pipe materials with fresh water did not produce any CN⁻ species. In light of these results, decontamination studies with bleach or the other detergent additive were not performed.

4.1.1.6 Mercuric Chloride

Adsorption experiments showed that mercuric chloride ($HgCl_2$) reacted with metal pipe materials. Metal oxides and chlorides were formed for all metals. For experiments using 36.9 g/L $HgCl_2$, amounts of mercury metal were formed for the brass and iron pipe samples and found in contaminant solution. Furthermore, a mercury mirror formed on the surface of the brass and iron pipe samples. The Cu substrate showed a light blue precipitate rather than evidence of mercury metal. However, SEM analysis showed droplets of mercury metal intermixed with copper oxide (Fig. 4.1.1.6.1).

Mercury concentration in solution was also monitored in adsorption experiments using 100 mg/L, 500 mg/L, and 1000 mg/L $HgCl_2$ tap water solutions with PVC and Cu pipe, respectively. The mercury concentration steadily decreased over exposure time for Cu pipe, but no significant changed was observed for PVC pipe (Fig. 4.1.1.6.2). These results further indicate a reaction with Cu, but not for PVC. Water characterization results for the $HgCl_2$/Cu experiments showed an increase in conductivity and turbidity, which increased with increasing initial $HgCl_2$ concentration. The TOC concentration for $HgCl_2$/Cu experiments only slightly increased compared to the initial tap water value. Conductivity and turbidity did not change significantly for $HgCl_2$/PVC experiments, but the TOC concentration was twice the value of tap water.

Given that mercury metal was formed with metal pipe materials, it was determined that decontamination would not be possible. PVC pipe material must be analyzed for the presence of mercury before determining if decontamination would be appropriate. A flushing of tap water with 5 % nitric acid would work to solubilize any mercury compounds.

4.1.2 Biological Contaminants

Results of bench scale tests of water mixtures with biofilms, spores, bacteria, and ricin are provided in Subsections 4.1.2.1 through 4.1.2.4.

4.1.2.1 Biofilm Reactor Measurements

The three major configurations for biofilm growth were the CDC bioreactor with baffle stirring (120 rpm), pipe section bioreactor with creeping flow (1 mL/min) and pipe section bioreactor with intermittent high flow (2.5 L/min, cycling at 2 h flow and 2 h with no flow). The biofilm organisms for these three biofilm reactor conditions reached similar levels during the three week growth period Fig. 4.1.2.1.1. These levels are comparable to the levels measured in a water distribution system that had measured values of 2.3×10^8 CFU/m^2 and 1.4×10^6 CFU/m^2 for PVC and copper, respectively (Schwartz et al., 2003).

4.1.2.2 Inactivation of Spores with Biofilm Conditioned Pipe Materials

The biofilm-conditioned coupons and pipe sections were then contacted with spores as described above. The CDC bioreactor was contacted with spores for 24 h with the stirring blade in place (180 rpm) and without stirring (only gentle mixing to prevent the settling of the spores). The second configuration with the pipe section with biofilm grown with creeping flow was contacted with BT spores in a beaker for 24 h (with gentle mixing to prevent the spores from settling). The third configuration with the pipe section reactor grown with high flow rate was contacted with BT spores for 24 h with intermittent high flow (2.5 L/min cycling at 2 h of flow and 2 h no flow). The concentrations of BT spores adhered to the different configuration are shown in Fig. 4.1.2.2.1. The configuration of the pipe section reactor with creeping flow (1 ml/min) and the CDC reactor without baffle stirring had low levels of adhesion. While the CDC reactor with baffle stirring (60 rpm) during contacting and the pipe section reactor with intermittent high flow (2.5 L/min) during contacting had much higher levels of adhered BT spores. This data shows the effect of mixing and high flow on increased concentrations of BT spores adhered to the biofilm-condition coupons and pipe sections.

In the disinfection experiments using the first two configurations, the coupons and pipe sections were contacted with chlorine and monochloramine solutions in containers without shear. In these experiments the spores adhered to the biofilm conditioned coupons, and pipe sections were very resistant to treatment by chlorine or monochloramine. Extremely high concentrations of chlorine (100 mg/L) were required to achieve any significant reduction in the levels of adhered spores (Morrow et al., 2008).

The disinfection results obtained with the third configuration using high flow rates were very different compared to the first two configurations. In the third configuration, the pipe sections were flushed with water and then chlorine solutions (10 mg/L) both steps at high flow rate. In these experiments a flow rate of 2.5 L/min in 19 mm diameter pipe sections, resulting in a Reynolds number of approximately 2800, indicating the flow is transitioning to turbulent. The effect of a 2 h flush with tap water on the removal of spores was reduction by approximately 0.5 $\log_{10}$.

Using the third configuration the chlorine disinfection was done using two solutions of chlorine. This was because the large surface area of the pipe sections and tubing with the biofilm resulted in a large chlorine demand. The first chorine solution (2 L with concentration of 10 mg/L) was depleted to approximately half of the initial value after 30 min. At this time a new chlorine solution was used to complete the disinfection process. The second disinfection solution (2 L with concentration of 10 mg/L) was also reduced to approximately half of the initial value by the end of the second disinfection step (150 min). The first 30 min disinfection step resulted in a reduction of approximately 1.5 $\log_{10}$ reduction and the second disinfection step resulted in a further reduction of approximately 1.5 $\log_{10}$ of BT spores associated with the surface The overall total reduction of BT spores was approximately 3 to 4 $\log_{10}$ in the pipe section loop for both the PVC and copper pipe sections (Fitzgerald et al., 2009). The biofilm organisms were reduced by a similar magnitude during the disinfection process using the high flow rate conditions. This is in contrast to the minor reduction of biofilm organisms seen in the disinfection steps using the first two configurations. These results prove the value of a high flow rate to achieve disinfection of BT spores adhered to biofilm conditioned pipes, when compared to previous studies using low flow rates.

Studies were also done using germinants to determine if germinating the spores would increase the effectiveness of the disinfection procedures. Biofilm conditioned coupons were prepared using the CDC reactor as previously described. The coupons were contacted with solutions of BT and BA spores (approximately 10^7 CFU/mL) for 24 h and then rinsed (Morrow and Cole, 2009). A germinant solution (1 mM inosine and 8 mM glycine) was then contacted with the spores adhered to the coupons for 24 h with stirring in the CDC reactor vessel. The effect of the germinants was to significantly decrease the concentration of the spores on the surface of the biofilm-conditioned coupons (reduction of 1.5 to 3 $\log_{10}$). The sensitivity of the germinated spores in solution and adhered to the biofilm conditioned surfaces to disinfection by chlorine, monochloramine, and heat (50 °C) was significantly increased (Morrow and Cole, 2009). These results indicate that the use of germinants is a promising approach to increase the effectiveness of disinfectant solutions, avoiding the use of high concentrations of chlorine that are damaging to the plumbing structure and harmful to the environment.

4.1.2.3 Measurement of Inactivation of Bacteria in Solution

The inactivation of spores and bacteria in solution were measured to have data to compare the performance of the inactivation when the spores or bacteria are adhered to biofilms on the surface of pipes. A common method to express inactivation of bacteria is to multiply the concentration of the disinfectant by the time required to achieve a 2 $\log_{10}$ or a 3 $\log_{10}$ reduction in the viable concentration of bacteria resulting in a CT value. A number of important solution variables will determine the CT values obtained. Temperature, pH, the ionic composition, and other components in the solution can have a major effect. CT values were measured using 10 mg/L chlorine in synthetic water (pH 8.2, ambient temperature) for BT and BA of 615 (13) mg min/L and 294 (12) mg min/L, respectively for a 2 $\log_{10}$ reduction (Morrow et al., 2008). One standard deviation of the mean is the value in parenthesis. The CT value of BT spores was measured in 0.1 M sodium phosphate buffer (pH 7.8) using 10 mg/L chlorine at 150 (60) min·mg/L 2 $\log_{10}$ reduction (Fitzgerald et al., 2009). The differences in these two values may be due to the lower pH of the phosphate buffered and possible stabilizing effect of components in the synthetic water.

4.1.2.4 Inactivation of Ricin by Disinfectants

The biological activity of ricin, lactate dehydrogenase, and lysozyme was measured after treatment with solutions of active chlorine and monochloramine. Active chlorine was very efficient inactivating all three proteins. Low concentrations of bleach (2.8 mg/L to 3.2 mg/L) added to ricin at a concentration of 67 μg/mL inactivated over 93 % of the biological activity in 10 min, and higher concentrations of bleach (5.6 mg/L) resulted in no detectable biological activity of ricin (Cole et al., 2008). The simulant proteins lactate dehydrogenase and lysozyme were also readily inactivated by chlorine and the reaction with monochloramine was much slower. Lactate dehydrogenase and lysozyme had a good correlation between the inactivation of biological activity and the loss of native fluorescence. Analysis of the proteins by gel electrophoresis and size exclusion chromatography indicated that the treatment with bleach resulted in extensive modification of the structure observed as modification of the protein charge and increased size due to aggregation.

41

The native fluorescence of proteins due mainly to trytophan and to a lesser extent tyrosine and phenylalanine was reduced by the treatment with chlorine. A good correlation between the reduction of biological activity and the decrease in native fluorescence in all three proteins was observed, indicating the utility of monitoring of fluorescence as a way to measure the inactivation of toxins such as ricin. Monochloramine solutions reacted with much slower kinetics with the proteins and was not effective at inactivating ricin even at levels 10-fold higher than chlorine (Cole et al., 2008). The similar inactivation kinetics with chlorine and monochloramine and the correlation between inactivation and loss of native fluorescence in the model enzymes (lactate dehydrogenase and lysozyme) make them useful simulants to study decontamination processes. Fluorescence is a real time measurement that can be done in a noninvasive manner and does not require additional reagents or complex sample manipulation.

4.2 Dynamic Fluid/Surface Interface Measurements

The dynamic fluid/surface contamination and flushing measurements that were made following the procedure described in Subsection 3.2 are described here. The contamination measurements over an approximate 200 h time period were made for three different Reynolds numbers varying from 0 to 7000:

$$\mathrm{Re} = \frac{4\dot{m}}{\mu_b P_w} \qquad\qquad (4.2.1)$$

where the wetted perimeter of the channel was 195 mm, the viscosity of the mixed bulk flow (μ_b) was calculated using a nonlinear mixture equation, and the mass flow rate ($\dot{m}$) was obtained from the turbine meter. Flushing measurements were done for a fixed Re of approximately 5000. The range of Reynolds numbers result from using a range of volume flow rates that a half-inch diameter tube would experience in typical buildings. After each contamination tests, the test surface was cleaned with acetone and clean tap water.

4.2.1 Contamination Excess Layer Thickness

Figure 4.2.1.1 provides the measured diesel layer thickness on a PVC surface as caused by an exposure to a flowing water/diesel (99.85/0.15) mixture, i.e., diesel fuel at approximately 0.15 % bulk mass fraction (1500 ppm). The exposure time is the duration of the exposure test: it is the time that the test surface is exposed to the contaminated flow starting with a clean surface. The open circle, square, and triangle symbols represent contamination measurements for the Re = 0, 3200, and 7000 conditions, respectively. Figure 4.2.1.1 shows that the Re = 3200 and the soak (Re = 0) contamination tests gave similar results. Specifically, the excess layer thickness was established immediately upon exposure of the PVC surface to the water/diesel mixture and remained nearly constant for the 200 h test duration. Only a marginal increase in the time-averaged l_e was observed from approximately 1.32 µm to 1.48 µm when the Re was increased from 0 to 3200, respectively. However, an increase in the Re to 7000 resulted in roughly a 142 % increase in the time-averaged l_e over the soak condition to an average value of approximately 3.2 µm. In addition, the Re = 7000 condition did not produce a nearly

42

constant l_e with respect to exposure time as did the Re = 0 and Re = 3200 conditions. Rather, the Re = 7000 condition gave a maximum diesel thickness of approximately 4.4 μm at an exposure time of 20 h. With further exposure to 140 h, the diesel thickness decreased from this maximum to nearly the thickness at the initial contamination, which was approximately 3 μm. For the PVC surface, the approximate average l_e for the Re = 0, 3200, and 7000 conditions was 1.32 μm, 1.48 μm, and 3.16 μm, respectively. Averaging over all contaminating flow rates and exposure times, the average l_e for x_b = 0.15 % on the PVC surface was approximately 2.0 μm.

Figure 4.2.1.2 gives the measured diesel layer thickness[5] on an iron surface due to exposure to a flowing water/diesel (99.85/0.15) mixture, i.e., the same composition as for the PVC surface. In general, the flow rate had little effect on the diesel excess layer thickness. For the iron surface, the approximate average l_e for the Re = 0, 3200, and 7000 conditions was 0.87 μm, 0.66 μm, and 0.60 μm, respectively. Averaging over all contaminating flow rates and exposure times, the average l_e for x_b = 0.15 % on the iron surface was approximately 0.71 μm. Consequently, the PVC surface adsorbs approximately 180 % more diesel fuel than the iron surface. The average accumulation of diesel fuel on a copper surface for x_b = 0.2 % was comparable to the PVC surface being approximately 2.3 μm (Kedzierski, 2006).

Figure 4.2.1.3 crossplots all of the contamination excess layer measurements of Fig. 4.2.1.1 as a function of Re. Figure 4.2.1.3 shows that the maximum diesel excess layer thickness on the PVC surface of approximately 4.4 μm occurred between Re of 5700 and 6300. The peak l_e for Re near 3200 was approximately 2 μm, which is approximately 55 % less than the maximum l_e for the Re = 6000 tests. Another 20 % reduction in the peak l_e on the PVC surface was observed when the nominal Re was reduced from 3200 to 0. The peak l_e for the soak tests (Re = 0) was approximately 1.6 μm for the PVC surface.

Figure 4.2.1.4 crossplots all of the excess layer measurements on the iron surface as a function of Re. Figure 4.2.1.4 shows that the maximum film thickness of approximately 1.3 μm occurred for the soak tests on the iron surface. The peak l_e on the iron surface decreases slightly from the soak condition for increasing Re. The peak l_e on the iron surface for Re near 3200 and 6000 was approximately 1.1 μm and 1.0 μm, respectively. The dashed lines given in Figs. 4.2.1.3 and 4.2.1.4 indicate the maximum measured excess layer for tests on the PVC and the iron surface. The variation in Re for a given set of tests for "fixed" Re was caused by an approximate 1 % variation in the water temperature during start-up and the an approximate 15 % variation in the water flow during the nearly 200 h test duration.

Figures 4.2.1.5 provides the measured diesel layer thickness on oxidized copper as caused by an exposure to a flowing water/diesel (99.8/0.2) mixture, i.e., diesel fuel at approximately 0.2 % bulk mass fraction (2000 ppm). The exposure time is the duration

---

[5] The soak measurements on the iron surface were corrected as outlined in Kedzierski (2008) to account for additional rust resulting when the surface was exposed to air during repair of the apparatus.

of exposure of the test surface to the flow starting from when the clean surface was first exposed to a particular flow condition. For all flow rates and exposure times, the average l_e for $x_b = 0.2$ % obtained from the eq. (6) methodology was approximately 2.3 μm.

Figures 4.2.1.6 provides the measured diesel layer thickness on oxidized copper as caused by an exposure to a flowing water/diesel (99.7/0.3) mixture, i.e., diesel fuel at approximately 0.3 % bulk mass fraction (3000 ppm). A much larger variability in the measurements is evident for the 0.3 % mass fraction than for the 0.2 % mass fraction condition. For all exposure times and Re, the average l_e for $x_b = 0.3$ % was approximately 7.4 μm, which is 5.1 μm (222 %) thicker than the average thickness observed for the 0.2 % mass fraction tests.

Figure 4.2.1.7 crossplots all of the excess layer measurements on oxidized copper of Fig. 4.2.1.5 as a function of Re. Figure 4.2.1.7 shows that the maximum diesel excess layer thickness of approximately 8 μm occurred at a Re near 4800. For Re larger and smaller than 4800, the diesel excess layer was thinner. For example, the l_e for Re near 1900 and 3800 was approximately 1 μm, which is nearly eight times less than the maximum l_e. The l_e for Re greater than 6000 was approximately 3 μm. Figure 4.2.1.8 crossplots all of the excess layer measurements of Fig. 4.2.1.6 (the 0.3 % mass fraction tests) as a function of Re. Figure 4.2.1.8 shows that the maximum film thickness of approximately 26 nm occurred at a Re of approximately 4000. Consequently, a maximum for the diesel adsorption exists near a Re of 4000 for both freestream concentrations. The dashed lines given in Figs. 4.2.1.7 and 4.2.1.8 represent the maximum measured excess layer for each range of Re tests. The variation in Re for a given set of tests for "fixed" Re was caused by an approximate 1 % variation in the water temperature during start-up and an approximate 15 % variation in the water flow during the nearly 200 h test duration.

4.2.3 Flushing Excess Layer Thickness
The flushing tests that were done after each contamination test are shown in Figs. 4.2.1.1 and 4.2.1.2. Measurements of l_e during flushing of the surface after the Re = 0, 3200, and 7000 contamination tests are represented by the filled circle, square, and triangle symbols, respectively. For the PVC surface, most of the flushing measurements are close to but less than zero. The average of all the flushing measurements on the PVC surface is approximately –0.2 μm. It is likely that an unknown bias error has caused the measurement to be less than zero because 0.2 μm is larger than the uncertainty of the l_e measurement. The negative thicknesses are interpreted as a clean surface. Consequently, the surface is clean nearly immediately after the inception of flushing. Negative thickness were observed with the exception of the flushing tests after the Re = 7000 contamination (of the PVC surface) where the initial l_e was about 0.6 μm. The l_e decreased from approximately 0.6 μm to approximately 0.13 μm after flushing for approximately 3.6 h. This corresponds roughly to a 0.13 μm/h removal rate, which is similar in magnitude to the flushing diesel removal rate, 0.10 μm/h, found for a copper surface. See Kedzierski (2006) and Fig. 4.2.1.5.

The l_e averaged over all exposure times and Re for the flushing test on the iron surface was roughly 0.44 mm. The flushing tests after the Re = 7000 contamination were the most successful in getting the surface clean ($\overline{l_e} = -0.2 \mu m$) assuming that negative thickness beyond the uncertainty implies a clean surface. On average, the flushing after the Re = 0 and the Re = 3200 tests left similar quantities of diesel fuel on the surface, 0.6 μm and 0.8 μm, respectively. These numbers indicated that flushing after the Re = 3200 contamination tests did not show any diesel fuel removal, while the flushing after the soak tests gave only roughly a 0.3 μm diesel fuel removal. However, both flushing tests after Re = 0 and the Re = 3200 contamination tests were corrected as outline in Kedzierski (2008). The corrected measurements have an estimated uncertainty of ± 0.8 mm. Consequently, it is likely that the iron surfaces were flushed clean after Re = 0 and the Re = 3200 contamination tests on the iron surface given that the measurements are within the measurement uncertainty and that the non-corrected flushing measurements indicated that the iron surface was clean after flushing.

Flushing tests done after the water/diesel (99.8/0.2) 4800-Re contamination tests on oxidized copper are shown in Fig. 4.2.1.5. The flushing measurements start at an l_e near 6.5 μm, which agrees with the value of l_e at the end of the 4600-Re contamination tests, thus, confirming the repeatability of the measurement technique. The l_e decreased from approximately 6.5 μm to approximately 1.5 μm after flushing for approximately 55 h. This corresponds roughly to a 0.09 μm/h removal rate and a 77 % reduction of the total diesel thickness over 55 h.

The flushing tests shown in Fig. 4.2.1.7 performed after the 5000-Re water/diesel (99.7/0.3) contamination tests on oxidized copper, likewise start at approximately the same l_e (1.5 μm) as where the previous contamination test ended, again demonstrating good repeatability. After approximately 20 h of flushing, the 5000-Re contaminant thickness was reduced to approximately –0.5 nm. Given the uncertainty of the measurement, most all of the diesel fuel has been removed from oxidized copper by flushing with clean tap water. The removal rate achieved after the 5000-Re, 0.3 % mass fraction (3000 ppm) contamination tests (0.1 μm/h) agrees closely with that achieved for the flushing tests done after the 4600-Re, 0.2 % mass fraction (2000 ppm) contamination tests. This suggests a constant removal rate of approximately 0.1 μm/h of diesel fuel from a copper surface for a flushing Re of 5000 that is independent of initial thickness and original contamination concentration. No removal rate could be calculated for the flushing tests done after the 7000-Re water/diesel (99.7/0.3) because the tests produced an l_e near –0.5 nm for nearly all measurement times.

4.3 Screening Tests
The detailed measurement results from the screening tests are presented in the following.

4.3.1 Diesel Fuel
Figure 4.3.1 presents measurement results for the diesel fuel contaminant. Diesel fuel is clearly seen to stick to each of the substrates, particularly the cast iron, PVC, and copper coupons. The initial one hour clean water flush was successful at reducing the diesel accumulation by about one order of magnitude. Continued flushing removed additional diesel, the removal rate slowed noticeably. Residual diesel fuel levels following 24 h of clean water flushing leveled off at about 4 % of the initial values.

4.3.2 Gasoline

Figure 4.3.2 presents the results of measurements for gasoline exposure. Gasoline accumulation after initial exposure was greatest for copper, and least for iron, with PVC and rubber falling in the middle. Flushing only was effective for the rubber substrate, which showed a reduction to approximately 3 % of the initial level.

4.3.3 Strychnine

Figure 4.3.3 presents the measurement results for the strychnine exposure tests. These results show that strychnine accumulated on all of the substrates, and flushing with tap water was not very effective in removing the residual contaminant.

4.3.4 Cyanide

Figure 4.3.4 presents measurement results for the cyanide exposure tests. The greatest accumulation was on the copper sample, which also showed some reduction from flushing. The other samples accumulated less cyanide, but the levels were not substantially reduced by flushing.

4.3.5 Phorate

Figure 4.3.5 presents the measurement results for the phorate exposure tests. Due to its low solubility in water, phorate accumulation was observed to be nonuniform, a probable explanation for the low initial measured accumulations on copper and PVC (i.e., measurement area at a spot where a lower amount of phorate was present. Flushing with water was not effective at removing the accumulated phorate from the PVC and copper, but performed better on the rubber and cast iron, although still leaving significant residual amounts.

Table 4.3.1 summarizes the results from the screening tests regarding whether the contaminant was observed to stick to the substrate and whether it was removed after flushing with water for a particular duration of time. Plots of the measurements along with a brief description of the data are provided in the subsections that follow.

Table 4.3.1 Results from Screening Tests			
Contaminant	**Substrate**	**Did it stick**	**Did it flush with water**
Diesel fuel	copper	Yes	Partially in 24 h
	iron	Yes	Partially in 24 h
	PVC	Yes	Partially in 24 h
	rubber	Yes	Partially in 24 h
Gasoline	copper	Yes	Not in 20 h
	iron	Yes	Not in 20 h
	PVC	Yes	Not in 20 h
	rubber	Yes	Partially in 24 h
Strychnine	copper	Yes	Not in 20 h
	iron	Yes	Not in 20 h
	PVC	Yes	Not in 20 h
	rubber	Yes	Not in 24 h
Cyanide	copper	Yes	Not in 24 h
	iron	Yes	Not in 24 h
	PVC	Yes	Not in 24 h
	rubber	Yes	Not in 24 h
Phorate	copper	Yes	Not in 24 h
	iron	Yes	Not in 24 h
	PVC	Yes	Not in 24 h
	rubber	Yes	Not in 24 h

As mentioned earlier, the Raman measurements do not provide absolute quantitative information about the magnitude of contaminant accumulation. To provide some estimates of contaminant layer thicknesses, static mass measurements were made using water and diesel fuel to associate the measured Raman intensities with film thicknesses. The water measurements represent dilute water/contaminant mixtures. The static measurements were made by comparing the mass of dry and wet tube samples before and after soaking them in liquid for 24 h.

The initial film thickness that was associated to zero flushing time was calculated from the measured surface area and the mass of diesel adsorbed to the pipe surfaces. For diesel fuel, the copper and PVC tube samples had an adsorbed film thickness of $56 \ \mu m \pm 10 \ \mu m$ and $47 \ \mu m \pm 10 \ \mu m$, respectively, while for water, the film thicknesses were $76 \ \mu m \pm 10 \ \mu m$ and $47 \ \mu m \pm 10 \ \mu m$. Using these values, the thicknesses of the diesel layer for copper at the beginning of the screening tests can be estimated to be 56 um, which corresponds to a measured Raman intensity of 100. As flushing proceeds, the relative change in Raman intensity can be used to estimate the thickness of the remaining diesel layer.

These results are plotted in Fig. 4.3.6 along with other measurement results obtained using an in situ fluorescence technique presented in Subsection 3.2. In those measurements, a diesel/water mixture was circulated until diesel accumulation reached a steady value, followed by flushing with clean tap water. It is interesting to note the similarity in initial measured film thicknesses for the two exposure techniques, although the static tests are somewhat higher. Removal of the accumulated diesel fuel is faster for the screening tests, undoubtedly due to flushing at a higher Reynolds number.

4.4 Hot Water Heater Tests

The measurements consistently showed that most if not all of the contaminants did stick to the tank surfaces and/or sediments after the initial exposure. Some contaminants, such as diesel fuel and toluene, showed a substantial reduction from flushing with clean tap water. Other contaminants, like phorate, gasoline, and the biologicals, required the addition of high levels of chlorine for effective removal. For the different measurements that were conducted, both exposure and flushing conditions varied considerably, but the results were surprisingly consistent regarding the tendency of the particular contaminants to stick, and the difficulty of removing them by flushing with tap water. It is not clear if water flushing alone can effectively remove strychnine, phorate or cyanide contamination, but even if it did, it is likely that the required water volumes would be so large as to be impractical. A better approach might be to flush with hot water, water with high chlorine levels or water with detergent. How this might be accomplished is discussed in Section 6. The details of the hot water heater tank measurements are as follows.

4.4.1 Diesel Fuel

Approximately 8 L of commercial diesel fuel were put into a previously used 189 L (50 gallon) hot water heater, which was then filled with cold water, resulting in a nominally 4 percent solution. That solution was circulated from the drain to the cold water inlet for 8 h to mix the solution and promote wetting of the tank interior and the sediment in the bottom of the tank by the diesel-water mixture. The tank was drained and a sediment sample extracted from the bottom of the tank. The tank was filled with cold water and allowed to stand for 4 weeks, at which time it was drained again, and another sediment sample was taken. The tank was re-filled with water, and flushed with clean cold water at 0.13 L/s (2 gpm) for a total of 24 h, corresponding to a water volume of approximately 10900 L (2880 gallons) being passed through the tank. The tank was then drained, refilled and flushed with hot water at 0.25 L/s (4 gpm) for 1-1/4 h, followed by another sediment sample extraction. To attempt to clean the sediment more thoroughly, the tank was drained, six cups of powdered laboratory detergent were put in, and the tank was filled with hot water and allowed to soak (no flushing) for 4 h and then drained before another sediment sample was taken.

Figure 4.4.1.1 shows the measured results for the testing sequence. The higher point to the left of the graph corresponds to the initial Raman spectrum for the sediment following the exposure. The lower point to the left of the graph was taken after the tank was drained and refilled, which showed little change in the Raman intensity. The data point to the right of the graph corresponds to a sediment sample taken after all of the flushing was completed. It is seen that the Raman intensity was somewhat lower than it originally had been but that it was only reduced to about 30 percent of the initial value. Following the cleaning operation, the sediment sample showed no characteristic Raman peak for diesel fuel.

These results suggest that simple draining and refilling of the hot water tank is not particularly effective at removing the diesel fuel, in large part because of the lack of solubility of diesel and its tendency to float on top of the water. A better approach might be to extract it from the top of the tank if possible, and avoid encouraging contact between the diesel fuel and the sediments at the bottom of the tank. While cold water flushing was not effective, cleaning with hot water and detergent was successful at removing all measurable traces of diesel fuel from the tank. In practice, this method would require injecting the cleaning solution upstream of the tank, or through the drain valve.

For this test, the tank heating elements were energized so that the water was heated to approximately 60 °C. A heated hot water tank was used to try to be more representative of actual operation, although the process of flushing introduces cold water to the tank in any case. The tank was partially drained to allow the addition of diesel fuel at a concentration of 4 % by volume, which was then mixed using a circulating pump, and allowed to stand for 24 h. This was followed by flushing with clean tap water at 0.25 L/s (4 gpm), while periodic water samples were taken from the tank outlet and drain for analysis. Because the tank was sealed, it was not possible to collect sediment samples. Figures 4.4.1.2 and 4.4.1.3 indicate that flushing was only partially effective at removing the diesel contamination from the tank as a significant amount of diesel remained present in water samples following 24 h of flushing. Another challenge associated with flushing hot water tanks is their large volume of water relative to flush water volumes, which causes water velocities in the tank and correspondingly low removal rates.

4.4.2 Strychnine
Raman spectra were obtained for the sediment samples at 0, 1, 3, and 24 h of flushing at 0.13 L/s (2 gpm). As shown in Fig. 4.4.2, strychnine was still present in sediment samples after 24 h of flushing, at about 10 % of the initial levels.

4.4.3 Sodium Cyanide
The sodium cyanide was detected in the initial sediment sample following exposure, but was not detected in subsequent samples. However, other cyanide measurements suggested that cyanide was not stable in chlorinated water and also reacted with substrates to form complexes that confounded measurements. So, a definitive conclusion could not be reached.

4.4.4 BT Spores
In this experiment, a concentrated solution of Bionide spores (an oil spore suspension) was used. 100 mL of concentrated Bionide and diluted it out to 3.78 L (1 gallon) in RO water containing 0.01% Triton X-100 and mixed well. The 189 L (50 gallon) hot water heater tank was filled one-half full and then the Bionide solution was added and the tank was filled up to the 189 L (50 gallon) mark. After allowing the tank to sit for 24 h, the tank was mixed with a rod and then flushed. After a total of 8 h of flushing (48 h after the spores were added) the sediment and water from the bottom of the tank was collected. To the sediment sample, 30 mL of PBS-Trition X100 was added. The anode, which was

immersed during the experiment, was scrapped with a cell scrapper and washed with 20 mL of PBS-Triton X-100.

The results showed the BT spore concentration remained the same in the hot water heater after it was allowed to sit for 24 h. In both trials, the spore concentration in the 10 min flush was similar to the spore concentration in the tank. This means the majority of the spores were flushed out in the 10 min flush sample. A 60 min sample was taken in Trial #1 and there was a 2 log decrease in the number of spores. However, the sample taken from Trial #1 after 2.5 h of flushing revealed no spores. In the second trial, a sample was taken after 10 minutes of flushing and the next sample was taken after 100 minutes of flushing. The concentration in the 10 min flush was similar to the initial concentration leading one to believe that spores were easily flushed out of the hot water heater. The next sample was taken at 100 minutes with no detectable spores in this flush sample as is shown in Fig. 4.4.4.1.

Even though a significant number of spores were flushed out of the hot water heater, a number remained in the sediment. After 24 h in Trial #1 there was the same concentration of spores in the sediment as what was reported in solution. After 48 h of contact in both experiments, there was a log decrease in the number of spores which resided in the residue. Additionally, there were some spores which remained on the anode (Trial #1) and sample of hot water tank wall (Trial #2) as seen in Fig. 4.4.4.2.

4.5 Full-Scale Tests
The measurement results for the full-scale plumbing loop were as follows.

4.5.1 Diesel Fuel
Diesel fuel was mixed with water at a nominal concentration of 8.3 %, then circulated through the pipe loop and allowed to stand for eight days. The water lines were then flushed at a flow rate of 0.25 L/s (4 gpm) at the pipe samples. Pipe samples were collected from Floor 1 (3/4 inch copper pipe) and Floor 3 (1/2 inch CPVC) and Raman spectra were obtained for all of the samples taken. The measurements showed that flushing with cold tap water was not very effective at removing the accumulated diesel fuel from the pipe samples for either the copper Fig. 4.5.1.1 or CPVC Fig. 4.5.1.2 pipe at this flow rate for flushing times of 13 h, although there was some reduction for the CPVC pipe.

4.5.2 Strychnine
Samples were measured for Floor 1 (3/4 inch copper pipe) and Floor 2 (1/2 inch copper pipe). For Floor 2, no Raman spectrum was obtained for the sample taken before flushing began and reasonable spectra were only obtained for the samples taken after approximately 1 h and 2 h of flushing: The measurement results are shown in Fig. 4.5.2.1 for ¾ inch copper pipe and Fig. 4.5.2.2 for ½ inch copper pipe, which indicate that cold water flushing was not effective at removing the accumulated strychnine from the pipe samples.

4.5.3 BT Spores
A solution containing BT spores was prepared in the following manner:
- Add 2 Liters RO water to 3.78 L (1 gallon) jug
- Add 4 mL 10 % Triton X-100, mix
- Mix bottle of bonide (#4) and add 100 ml to jug and mix well
- Add 1.67 L RO water to jug and mix well again
- Titer jug

This solution was circulated through the pipe test loop and left overnight to soak. This was followed by flushing and sampling at 1 h, 3 h, and 13.3 h.

Pipe samples were analyzed in the following manner:
- Received samples and sampled the next day
- Remove pipe from tube and dip pipe in beaker with 100mL RO water
- Repeat 2 more times using a new beaker with 100mL RO water
- Scrape pipes using cell scraper into 5 mL PBS/Triton in 50 ml conical tubes
- Wash pipe 4 times with buffer in tube
- Vortex for 30 sec
- Titer

The results of these measurements are shown in Figs. 4.5.3.1, 4.5.3.2 and 4.5.3.3 for the three floors, which had different pipe sizes and materials. The ¾ inch copper pipe Fig. 4.5.3.1 showed less than a one log reduction after 3 h of flushing, and a three log reduction after 13.3 h of flushing. While this is a significant reduction, it is possible that the remaining spores could germinate and eventually reproduce, causing them to persist in the system.

Figure 4.5.3.2 shows the results for ½ inch copper pipe, which showed a 2.5 log reduction after a 3 h flush. The 13.3 h flush sample could not be analyzed.

Figure 4.5.3.3 shows the results for the ½ inch CPVC pipe, which showed a 2 log reduction after 3 h of flushing, and a three log reduction after 13.3 h.
While substantial reductions in the number of spores were observed, complete eradication would probably require additional decontamination procedures, such as hot water, shock chlorination and/or germination.

Table 4.5.1 summarizes the results for the full-scale plumbing system tests for diesel fuel, strychnine, BT spores and cyanide. The results showed diesel fuel, strychnine and BT spores adhered to all of the surfaces of this study. Both diesel fuel and BT spores could be partially removed by flushing after approximately 12 h. However, strychnine was not removed in 14 h of flushing.

Table 4.5.1 *Results from Full Scale Tests*			
Contaminant	**Substrate**	**Did it stick**	**Did it flush with water**
Diesel fuel	copper	Yes	Partially in 12 h
	CPVC	Yes	Partially in 12 h
Strychnine	copper	Yes	Not in 14 h
	PVC	Yes	Not in 14 h
BT Spores	copper	Yes	Partially in 13 h
	PVC	Yes	Partially in 13 h
Cyanide	copper	NA	NA
	PVC	NA	NA

5.0 MODEL DEVELOPMENT

5.1 Semi-empirical

The derivations of the models to predict the thickness of the contaminant excess layer on plumbing surfaces for the contamination and the flushing conditions are presented in Kedzierski (2008) and reproduced here. Each model assumes that transfer of mass to and from the surface occurs solely in a direction that is perpendicular to the surface. The flushing model predicts the thickness of the excess layer as a function of time, contaminant transport properties, and flushing Reynolds number. The contamination model gives the maximum contaminant excess layer thickness that can occur for a given contaminant bulk mass fraction and surface affinity.

5.1.1 Flushing Model

The model for flushing with contaminant-free water is based on the conservation of contaminant mass within the excess layer (l_e). Because the excess layer is thin, it is approximated as stagnate in the axial direction such that the net motion of diesel fuel is solely perpendicular from the surface in the y-direction. The model is governed by turbulent convection and diffusion of contaminant from the surface. The effect of diffusion is modeled by the ratio of the diffusion coefficient (D_{wd}) to the transition depth (B_T) over which the concentration difference occurs. The turbulent convection is modeled by the contaminant viscosity (ν_d), the friction velocity (u_*), and an entrainment constant (K_J) that relates the average entrainment velocity ($v'_{max} / 2$) to the local axial-velocity (u) in the viscous sublayer as:

$$K_J \equiv v'_{max} / u \qquad (5.1.1.1)$$

The resulting equation to predict the contaminant excess layer (l_e) as a function of flushing time (t) and initial excess layer thickness (l_{e0}) is:

$$\frac{l_e}{l_{e0}} = \left(1 + \frac{1}{K_D}\right) e^{\frac{-K_J u_*^2}{2\nu_d} t} - \frac{1}{K_D} \qquad (5.1.1.2)$$

52

where the dimensionless constant K_D is a ratio of the convective to the diffusive influences:

$$K_D = \frac{K_J u_*^2 B_T l_{e0}}{2 v_d D_{dw}}$$

(5.1.1.3)

For $K_D = 1$, convection and diffusion fluxes of contaminant from the surface are equal at the beginning of the flushing. Values of K_D larger than 1 indicate that convection is more important than diffusion of contaminant from the surface.

The friction velocity is calculated from an equation given by Kays and Crawford (1980) with the average (bulk) axial velocity (and fluid properties) of the flushing water (V) and its Reynolds number as:

$$u_* = V \sqrt{0.039 \, \text{Re}^{-0.25}}$$

(5.1.1.4)

Flushing measurements for contamination levels larger than those of the scope of the present project were taken in order to establish changes in the measured excess layers that were large enough to fit to eq. (5.1.1.2). These experiments flushed diesel fuel from a PVC surface with tap water flowing at an Re of approximately 5000. The initial diesel excess layer was approximately 31.4 µm. Figure 5.1.1.1 shows the flushing measurements that were used to obtain the K_J and D_{dw}/B_T constants from a least squares regression of the Fig. 5.1.1.1 data to eq. (5.1.1.2). The regression constants are:

$$K_J = 0.66 \times 10^{-8} \pm 0.05 \times 10^{-8}$$

(5.1.1.5)

$$\frac{D_{dw}}{B_T} = 0.6 \times 10^{-10} \left[\frac{m}{s} \right] \pm 0.1 \times 10^{-10} \left[\frac{m}{s} \right]$$

(5.1.1.6)

Figure 5.1.1.1 plots eq. (5.1.1.2) using the regressed eq. (5.1.1.5) and eq. (5.1.1.6) constants. The model predictions are within ± 4 µm of all measured l_e. This translates into predictions being within 12 % of the measurements at the beginning of the flushing tests and more than 100 % larger than measurements near the end of flushing. This may seem to be an unacceptably large prediction error for the film thickness; however, the primary goal of the model is to predict flushing times to clean. To put it in perspective, the more than 100 % error in the prediction of the film thickness results in an overprediction of the flushing time by approximately 9 h, which is an acceptable error considering that a safety factor of at least two would be applied adding more than 50 h to the required total flushing time.

The entrainment constant, K_J, is expected to be less of a function of the properties of the contaminant/flushing pair than is the diffusion constant. The ratio of the axial bulk velocity to its peak fluctuating component may be nearly constant because the bulk velocity is the potential for the fluctuating velocity. Considering this and that Kays and

Crawford (1980) show that the ratio of the peak fluctuation turbulent velocity components is constant, the K_J may be relatively constant for a particular flushing fluid and for various Re. It is expected that the K_J presented here would be valid for water and liquids with similar kinematic viscosity.[6] Conversely, the D_{dw}/B_T depends on the properties of both the contaminant and the flushing fluid.

The required time (t) to flush the surface clean is derived in Kedzierski (2008) and is presented here:

$$t = \frac{-2\nu_d}{K_J u_*^2} \ln\left[\frac{1}{K_D + 1}\right] \qquad (5.1.1.7)$$

Table 5.1.1.1 compares predictions using eq. (5.1.1.7) with observed values. All of the flushing times are predicted to within 7 h and all but one prediction is conservatively overestimated.

Table 5.1.1.1 *Comparison of flushing model predictions to measurements*					
Surface	**Re**	l_{e0} **(μm)**	K_D	**Flushing time for clean eq. (5.1.1.7) (s)**	**Observed flushing time for no change (s)**
PVC	0	1.5	0.29	6.4	≈ 0
PVC	3200	1.5	0.29	6.4	≈ 0
PVC	7000	2.5	0.49	9.7	≈ 8
Iron	0	1.0	0.20	4.3	≈ 0
Iron	3200	1.0	0.20	4.3	≈ 0
Iron	7000	0.5	0.10	2.3	≈ 5

5.1.2 Contamination Model

For the contamination case, where a balance between deposition of contaminant on the surface and removal of the contaminant from the surface must be achieved, surface adhesion requires that a distinction be made between the velocity of the contaminant toward the surface (v_i) and that away from it (v_o). Very near the wall there is an additional resistance to flow away from the surface as caused by the attraction of diesel fuel molecules to the molecules of the pipe surface. Likewise, for the same region near the pipe, the attractive forces induce a reduction in the resistance of flow toward the surface. Considering that it is flow that is being modeled, a simple way to approximate this behavior is via a modified viscosity. For example, the entrainment velocity (evaluated using the properties of the contaminated flow) given in Kedzierski (2008) for flow approaching the pipe surface becomes:

$$\left[v_i\right]_{y=l_e} = \frac{K_J u_*^2 l_e}{2\nu_d \tanh\left(\dfrac{l_e}{\delta}\right)} \qquad (5.1.1.8)$$

---

[6] However, K_J may be altered by the adhesion forces that must be overcome to remove the contaminant from the wall.

Here, the kinematic viscosity is less than the constant property viscosity (ν_d) for distances from the surface less than the penetration depth, δ, because adhesive forces assist the flow of contaminant to the surface within this region. The magnitude of the penetration depth is determined by the affinity of the contaminant for the pipe surface. Consequently, each contaminant\pipe combination will have its own value of δ. The hyperbolic tangent was chosen for its simplicity and because it closely matched what was believed to be the required relationship with respect to l_e.

Likewise, surface adhesion acts to deter the flow of contaminant from the wall according to an assumed hyperbolic cotangent relationship with respect to l_e. Here the leaving velocity is approximated as a local increase in viscosity above the constant property value as the pipe surface is approached for distances less than δ:

$$[v_o]_{y=l_e} = \frac{K_j u_*^2 l_e}{2\nu_d \coth\left(\dfrac{l_e}{\delta}\right)} \tag{5.1.1.9}$$

Figure 5.1.2.1 provides an example of the ratio of the adhesive influenced viscosity to the constant property viscosity of diesel fuel (ϑ) as a function of the distance from the iron surface. For this example, the excess layer is 0.5 μm and the penetration depth is 14 μm. The viscosity ratio is plotted as a function of y to illustrate the waning and waxing influences of the adhesive forces between iron and diesel fuel as a model concept. However, as far as the contamination model is concerned, ϑ is evaluated only at $l_e = 0.5$ μm.

Considering that the exposure time of a pipe surface to a contaminant may not always be known, a conservative decontamination response would be to flush for the maximum contaminant film thickness (excess layer) for the given concentration of contaminant in the flow. The maximum contaminant excess layer, which occurs at steady state where the contaminant deposition balances the removal at the wall, can be determined by setting the partial derivative of the excess layer with respect to time to zero (see Kedzierski, 2008) and solving the resulting equation for l_e, yielding:

$$[l_e]_{max} = \frac{\delta}{2} \ln\left(\frac{1+\sqrt{x_b}}{1-\sqrt{x_b}}\right) \tag{5.1.1.10}$$

Equation (5.1.1.10) can be used to determine the maximum possible contamination level and used as input to eq. (5.1.1.7) in order to calculate the required flushing time to obtain a clean surface. Mathematically, eq. (5.1.1.10) is valid for values of x_b between zero and 1; however, it has been validated for only dilute solutions.

Average values of the penetration depth were found by back-substituting the measured diesel bulk mass fraction and the measured $[l_e]_{max}$ into eq. (5.1.1.10) and solving for δ. The δ was found to be surface dependent: 150 µm, 42 µm, and 16 µm for copper, PVC, and iron, respectively. Figure 5.1.2.2 shows the maximum contamination layer for the three pipe surfaces as a function of x_b as predicted by eq. (5.1.1.10). Because of the presumed stronger affinity between copper and diesel, the copper surface has greater contamination levels for a given x_b as compared to the PVC and iron surfaces. The greater affinity is modeled via the larger penetration depth.

Figure 5.1.2.3 is a preliminary validation of eq. (5.1.1.10) and its theory in that it compares well against independent measurements and exhibits an exponential decay with respect to flushing time. Figure 5.1.2.3 demonstrates the agreement between eq. (5.1.1.10) predictions, as shown by the solid lines, and two different measurement techniques. Raman intensity measurements from the technique in Subsection 3.3 are represented by the filled circle, triangle, and square. The second technique was the traversing fluorescence technique provided in Subsection 3.2 and depicted by open circles. The Raman screening tests were for copper, PVC and cast iron at Re=30,000. The fluorescence tests were for PVC at Re=5000. Removal at Re=5000 was slower than at the higher Re, and the time required to remove most of the diesel fuel was on the order of 50 h. Good agreement between the measured flushing times and the model was achieved. Each prediction coincides with measurements from both techniques for at least one value above zero. In addition, each prediction falls on zero between measured points that indicate a clean surface. It is not likely to measure the surface at the precise time that it has become clean. Consequently, it is not likely to have a measured point coincide with the model at zero film thickness.

5.1.3 Future Work on Semi-empirical Model
The overall goal of this project is to develop the necessary tools that are to adequately respond to a contamination event. These tools have been envisioned to consist of measured data, predictive models, and a computer program that embodies the data and models. The software would allow the user to input information that is specific to a contamination event and to receive estimates of contamination levels and required flushing times. Such an all-encompassing goal requires a significantly large effort to ensure that the software makes reliable predictions for all possible contaminants. To achieve this task in a reasonably short time, the model must be physically based, but yet simple, so that it may be easily adjusted and extrapolated for different contaminants.

In this light, the models developed in this project require additional enhancement to ensure that the best possible predictions can be provided. Possibly, the major effort toward this end will be using measured data to calculate the constants K_J, D_{dw}/B_T, and δ for potential contaminants like toluene and pipe surfaces. See Kedzierski (2008). In the course of this effort, it will be determined whether or not K_J is merely a function of the properties of the flushing fluid. In addition, it can be determined if δ is constant for a particular contaminant/surface pair. In the short term, more flushing measurements at different initial l_e are required to validate the flushing constants considering that only one data set was used to obtain them. It may be appropriate to lump the diffusion that occurs during contamination into the adhesion effect. On the other hand, inclusion of a diffusion component in the contamination model may be necessary. These questions can be answered with further experimentation and model enhancement.

5.1.4 Software to Predict Contamination Levels and Required Flushing Times

Computer software was developed to predict contaminant levels and required flushing times. The flushing and contamination models developed in Subsections 5.1.1 and 5.1.2, respectively, served as the basis of the computer program. The general concept is based on modeling the interaction between the contaminant and the pipe walls under different flow conditions to predict the contaminant accumulation and removal. Certain parameters are used as inputs to the software program, such as mass fraction of contaminant and exposure time, flushing Reynolds number, and the tool provides as output a recommended flushing time. Currently, the software tool has been validated using detailed measurements of diesel fuel contaminant, but it is under development to expand its capabilities to cover a wider range of contaminants. The complete computer software may be downloaded from: ftp://ftp.nist.gov/pub/bfrl/squid/.

Figure 5.1.4.1 shows an example of Graphical User's Interface (GUI) that is used to input the pipe material, flushing flow rate (Re), pipe diameter, contaminant, and other information relative to the contaminant incident. The GUI also shows the output in terms of the contaminant thickness on the pipe wall as a function of flushing time. Overall flushing time and estimated maximum contaminant thickness is also given.

In Fig. 5.1.4.2, the flushing model given in Subsection 5.1.1 illustrates the effect of the building floor area on the required flushing times. It was assumed that only a single entrance and a single exit were used to flush the building plumbing. As a result, piping downstream of the entrance is flushed with water that has entrained a small amount of contaminant in the flow. The contaminated flushing fluid may possibly redeposit on the surface or reduce the effectiveness of flushing due to a loss in concentration gradients for diffusion. For such an assumption, the required flushing time should increase with the building area because of effects like re-contamination of the surface and reduction in diffusion concentration gradients.

The total length of plumbing (L_T) in a building with floor area, A, was calculated based on the following, which was derived from a small construction data set:

$$L_T = 8\sqrt{A} \qquad\qquad (5.1.4.1)$$

57

Only the effect of diminished concentration gradient on the contaminant removal was modeled because it was believed to be the most significant effect. The mass fraction of the flushing fluid leaving the building was calculated from a Lagrangian control volume analysis on a fluid element as it progressed from the pipe entrance to its exit:

$$x_b = \cfrac{L_T}{L_T + \cfrac{VB_T\pi D^2 \rho_w}{4(K_D+1)D_{wd}\rho_d}} \tag{5.1.4.2}$$

The average contaminant mass fraction ($0.5x_b$) was used to correct the diffusion term of the model given in section 5.1.1 for the loss in diffusion as:

$$K_d = \frac{K_d}{(1-\overline{x}_b)} \tag{5.1.4.3}$$

Figure 5.1.4.2 shows the effect of the building area for three flushing Reynolds numbers. In general, as the Reynolds number increases, the effect of building area is diminished. In fact, for turbulent Reynolds number, the building area has a relatively negligible effect on flushing time. Only when the Reynolds number is reduced to 3 is a significant change in flushing time seen with increased building area. For the small flowrate, the diffusion of contaminant has become of the same order as is convection in terms of contaminant removal. Otherwise, turbulent flushing ensures that convective entrainment of the contaminant dominates resulting in little effect of building area.

5.2 Flow in idealized pipe geometries

Evaluation of the quality and safety of drinking water is of paramount importance for any community. When a pipe or water conduit system is contaminated, it is important to determine the best strategies for removal of contaminants in a timely fashion. The traditional approach for decontaminating plumbing systems is to flush pipes with water at high volumetric flow rates, primarily because this methodology is most easily implemented, and there are few simple alternatives. However, the effectiveness of this approach has not been demonstrated. In fact, for several reasons, it may not be the best way to remove contaminants. For example, it can be argued that because high velocity water flow in piping systems are in the turbulent regime, eddies that are generated tend to inhibit the transport and removal of contaminants. Further, it is not understood, to what degree, the location of the contaminant in the pipe system will affect its removal rate. For example, upon initial inspection of water, post flushing, it may appear that the contaminant had been completely removed only to find out that the contaminant reappears in the pipe system at a later time because it was located outside the main flow path. That is, the contaminant may have entered the main flow path at a later time via diffusive transport. We investigate these possibilities by comparing transport in different flow geometries at low (laminar) and high Reynolds numbers using numerical modeling. An effort is made to separate the effect of overall flow rate from the actual flow path the contaminant takes, which will change with Reynolds number. "Patches" of contaminant, are placed in different locations of the pipe system to help understand the effect of the local flow field on contaminant removal. We will also consider the effect of pipe length on decontamination times.

It should be pointed out that we cannot currently predict the actual efficiency of contaminant removal for any particular pipe component. This would entail detailed flow modeling for that specific pipe geometry with a correct understanding of the contaminant interaction with the pipe wall. In addition, the approach used in this study needs further experimental validation. Hence, the results presented in this work should be thought of, more so, as qualitative in nature with the hope of identifying potential problem areas in contaminant removal.

5.2.1 Flow Around Bends, Past Cavities, and Over Obstructions

To investigate the effect of pipe geometry on the movement of contaminants, four idealized flow geometries that could represent typical pipe system components, are considered: The base system for comparison to is a straight rectangular square pipe. The second system studied is a U shaped bend Fig. 5.2.1. This geometry could, for example, be representative of a valve in the pipe system. The third pipe system has a cavity or pocket (something like a shunt or trap) out of the line of the main flow. See Fig. 5.2.3. The fourth pipe geometry is a simple rectilinear obstruction in the flow path See Fig. 5.2.5. This geometry could be similar to a valve or an obstruction in the main flow path.

The computational fluid dynamics approach used in this study is based on a finite difference approximation of the Navier Stokes equations and is fully described in Martys (2001). This simulation approach has been validated for a variety of flow scenarios ranging from simple Poiseuille flow to obtaining correct solutions for boundary layer flow. For this study, the Navier-Stokes equation is used to generate flow fields in the different pipe geometries. The flow fields are then used to determine the contaminant transport via numerical solution of the advection diffusion equation Martys (1994). For this report, two Re values, 30 and 3000, which are representative of typical slow and fast flow rates in pipe systems respectively were considered. The Reynolds number, (Re), is given by $Re = <u> l / \nu$ where $<u>$ is the average velocity at the inlet, l is the channel width, and ν is the kinematic viscosity. In these simulations, Re is adjusted by varying the kinematic viscosity while fixing the flow velocity at the inlet. That way effect of the flow trajectory from the effect of simply increasing the flow velocity to increase the Reynolds number can be isolated. Figures. 5.2.1 and 5.2.2 illustrate the main flow features obtained from the simulations for the U shaped pipe and cavity geometries (Fig. 5.2.3), respectively. In both figures, the fluid enters from the left side and exits out the right side. As expected, the flow fields at low and high Reynolds number were quite different. At the lower Re (Fig.5.2.1), the flows appeared laminar with little rotational flow. In the case of the U shaped pipe, at higher Re (Fig. 5.2.2), significant rotational flow developed, especially around the pipe's corners. In the case of flow past a cavity, at low Re, a single rotational flow field developed near the opening of the cavity (not shown) but at the higher Re (Fig. 5.2.3) two counter rotational flows developed. In general, the higher Re flow typically develops a greater velocity near the surface along "straight sections" of pipe, as compared to the lower Re flow.

59

5.2.2 Contamination Model

The contaminant is treated as a scalar field whose time evolution is controlled by the advection diffusion equation in the limit of very small diffusion. The transport of the contaminant is considered passive. That is the contaminant does not affect the flow field. Contaminants are introduced into the pipe system either as a rectangular patch along the pipe surface or at the inlet as a constant source. The egress time for different flow scenarios is then compared.

In the case of the contaminant patch, a simple linear coupling of contaminant with the pipe wall where the rate of absorption is proportional to the concentration difference in the fluid and on the wall is assumed.
Here,

$$\frac{dc}{dt} = -k(c - c_s),$$

(5.2.2.1)

where, k is a is a kinetic reaction constant and c_s is the equilibrium concentration on surface. The simulation model is not limited to this surface interaction but we use it as a first approximation.

This work tries to separate the effect of overall flow rate to actual flow path the contaminant takes which will change due to Reynolds number. It is found that the Reynolds number can have a dramatic effect on the trajectory of the contaminant. At higher Reynolds number the flow near the boundary of straight paths is much greater allowing for more contaminant to be carried away into the main flow streams. Higher vorticity can be near some corners so that the contaminant has a tendency to stay in that region. Contaminants starting from a corner but closer to the outlet took longer to egress in the main flow path..

Figure 5.2.4 compares the amount of contaminant as a function of time for the different flow geometries and at different Re. The variable, C, is the total contaminant in the pipe system and is normalized by its initial value, $C_{init.}$ The black curves correspond to the straight pipe flow. The blue (Re = 3000) and green (Re = 30) curves correspond the case of flow in the U shaped region. The dashed lines correspond to the placement of contaminant in the region near the bottom right hand corner of the U shaped pipe, whereas the solid line correspond to a contaminant being placed at the midpoint of the pipe along the bottom. In general, placement of the contaminant away from the main flow path can make the egress time longer. In the case of the higher Reynolds number, the delay of removal is enhanced because of the slow circulatory flow in that corner. Clearly, at both Re, the contaminant left the pipe system much more efficiently when placed in the bottom straight section. When the contaminant is placed in the corner, a significant increase in time for its removal is due to the lower flow rates. In the case of the higher Reynolds number, the delay of removal is enhanced because of the slow circulatory flow in that corner.

The violet lines correspond to contaminant removal from the cavity. In this case, the contaminant was initially placed along the bottom of the cavity. As can be seen, the contaminant remains, for a significant time, in the cavity. Here the slow rotational flow inside the cavity has a tendency to suppress the contaminant's egress. At either Reynolds number the cavity was the most difficult to clean and presents the biggest challenge as the time scale for contaminant removal was significantly higher.

In Figs. 5.2.5 and 5.2.6 the cases of low and high Re, respectively, flow past a rectilinear obstruction. At the lowest flow rates the fluid simply moves up, over, and down the back of the obstruction. At high Reynolds number a vortex was created down stream from the obstruction. For the case of flow past an obstruction (Fig. 5.2.7) the relatively higher flow near the boundary at higher Re caused the contaminant to be removed more quickly. So the relative efficiency of removal is dependent on geometry.

5.2.3 Ingress of Contaminant
Although the placement of contaminant plays an important role in its removal, it rates a second consideration in the history of how the contaminant actually enters and passes through the pipe. In the Fig 5.2.8 the time evolution of a contaminant entering U shaped pipe system shown earlier for the Re = 3000 case. The visualization shows that regions which take longest to remove contaminant are also the same regions that take longest to reach from the inlet (see Fig. 5.2.8). In other words, locations that take longer to decontaminate are also less accessible to contamination. Of course it is important to note that only cases where there is hydrodynamic controlled motion of the contaminant is considered. If over time the contaminant enters a slow flow region by diffusion and is attached by say Van der Waals forces as in a bacterial agent, the removal time could be significantly longer.

5.2.4 Effect of Pipe Length
The effect of pipe length on contaminant removal by determining the time it takes a certain percentage of contaminant to leave a rectangular pipe as a function of the distance of the contaminant patch from the outlet was studied. Figure 5.2.9 shows for over a decade in relative distance of contaminant placement from the outlet and at relative remaining concentrations of 5 % and higher, the time it takes to reach a specified concentration scales linearly with distance of contaminant patch to the outlet. For lower concentrations, the time to reach the specified concentration level initially increased more rapidly with pipe length. However, it was found that while the linear scaling with pipe length does not appear at the shorter distances, it does trend to a linear relation approaching a similar slope with longer distance.

5.2.5 Effect of Flow Rates
While it was found that the flow fields associated with higher Reynolds number can strongly affect the trajectory of the contaminant it could be temporarily confined to regions of higher vorticity. Note, the actual flow rates needed to achieve the higher Re will, in general, more than compensate for the reduced efficiency of removal. For example, consider the U shaped pipe study. The contaminant placed near the corner (CP), when Re = 3000, took nearly 40 % longer in time to reach the same concentration then the Re = 30 case. However, these Re numbers were achieved by adjusting the viscosity so that the average flow rate was the same for those two cases. But, in real applications Fig. 5.2.10 shows the viscosity is fixed, and the flow rate would have to increase by a factor

of 100 to reach the same Re value. As a consequence, the time to reach the same level of decontamination was about a factor of 70 times faster at the higher Re. Another consideration is that the higher Re flow will require more water to reach the same level of decontamination.

It is clear that flow rates dominate the removal of contaminant over time. Delays in contaminant removal, due to eddies or vortex motion in flow, do not have as big effect as actually increasing the flow speed in the cases studied.
`
To further help clarify the effect of flow rate vs. geometry and contaminant placement, Table 5.2.1 gives a relative comparison of contaminant removal times.

Table 5.2.1	Comparison of Contaminant Removal Corrected for Flow Rates			
Re Number	**Straight Pipe**	**U-Shape (MP)**	**U-Shape (CP)**	**Cavity**
30	1	1.5	2.5	10
3000	0.009	0.012	0.03	0.18

Results are normalized to the case of the straight pipe at Re = 30. Table 5.2.1 shows, for example, the removal of the contaminant patch located at MP in the U-Shaped pipe at Re 30. This took 1.5 times as long to reach the same concentration as a similar patch in the straight pipe for the Re 30. But at Re 3000, removal of the same contaminant patch in the U-Shape pipe took .012 times as long as the Re = 30 Straight pipe case.

Note that in the case of high Re flow in the straight portions of pipe or the near the MP of the U shaped pipe, the removal of contaminant is slightly enhanced. In contrast, the removal of contaminants near corners or in a cavity, i.e. regions away from the main flow path, is less efficient at high Re.

The model development for flow in idealized pipe geometries has shown that pipe geometry, the location of contaminants, and flushing rates can have a significant effect on the decontamination of pipe systems. The higher the flush rate, the faster the removal of contaminant; however, in some cases, the effect of the vortices and dead flow regions could reduce the efficiency of removal. That is it will take a greater amount of water to reach the same contamination level when using high flush rates because the higher Re flow could have a significant effect on the kinematical motion of contaminants. In general, if only hydrodynamics is considered, parts of a pipe system that are hard to decontaminate are generally less likely to be reached by contaminates. However, this analysis does not when taken into account diffusion and mechanisms of wall interactions that are clearly important and could significantly affect removal times. Further study, such as, validation of the simulation model by close comparison with experiment is still needed to make this approach a predictive tool for contaminant removal.

6.0 DISCUSSION OF RESULTS
6.1 Inactivation and Flushing of Spores, Bacteria, and Ricin

General recommendations for the decontamination of biological threats in water systems are difficult to formulate because of the different conditions encountered in different systems and lack of data in real life systems. The age of the plumbing system and local water conditions will determine the type and amount of biofilm as well as corrosion present on the pipes. Corroded plumbing surfaces have been found to increase the persistence and susceptibility to disinfectants of bacillus spores (Szabo et al., 2007) and bacteria (Szabo et al., 2006) in model water systems.

Data on the inactivation measurements of biological threats in solution by chlorine using carefully controlled laboratory settings have been published in Rice et al. (2005), Rose et al. (2005), and Hosni et al. (2009). This data is valuable to design disinfection process, but models must take into account the different conditions that may be present in real world systems. Temperature, pH, solution ionic composition, presence of biologics in the water will all affect the disinfection process. A large safety factor for disinfection will have to be built into the design of any disinfection process. The properties of the biological materials will influence their behavior with disinfectants. For instance, when using chemical disinfectants such as chlorine, it can react with many contaminants present in the biological threat samples and naturally occurring materials in the water. The amount of chlorine reactive compounds present that chemically exhaust the active chlorine is termed the chlorine demand. The amount of active chlorine available to disinfect the biological threats is that added above the chlorine chemical demand level. The presence of biofilms and corrosion present in the water system will increase the difficulty in achieving total disinfection of the water system.

Due to the complexities and the uncertainties associated with the conditions and nature of the biological threats to be encountered in the real world, methods to monitor the actual disinfection process in real time are especially valuable. Monitors to measure chlorine demand and active chlorine in a disinfection process are valuable to provide feedback. Active chlorine levels can be adjusted as needed, assuring that an active chlorine level is maintained avoiding excessive amounts of chlorine added to the environment. Monitoring the decrease in native fluorescence for the inactivation of spores (Alimova et al., 2005) and protein toxins (Cole et al., 2008) also provides real time feedback about the state of the biological target. These processes have been studied in laboratory conditions but not with real water systems. It will be important to prove the utility of monitors with real world systems.

The use of chlorine dioxide found to be a more effective disinfectant of *Bacillus globigii* spores in solution (Hosni et al., 2009), and should be investigated for the disinfection of spores associated with biofilms in water systems. Two promising approaches, high flow rate and the use of germinants, should also be tested in large-scale systems using real world conditions. The use of high flow increased the effectiveness of disinfection in the pipe section bioreactors. Presumably the high flow in the pipe section reactor results in turbulent flow that results in an increased permeation of the biofilm by chlorine and mechanical scouring of the pipe walls. Achieving turbulent flow in small diameter pipes and all sections of a building water system may be difficult to achieve in some configurations.

The use of germinants is also a promising approach that does not involve the use of environmentally damaging chemicals. The germinants used in this research were highly purified reagent grade chemicals, alanine and inosine. It should be investigated to determine if low cost bulk chemicals, such as food supplements may work as well as the more expensive reagent grade chemical in a large-scale water system. The use of biological stimulants determined in the project should be tested for disinfection efficiency in the large-scale system to determine the efficiency of disinfection in a near to real life system. As in the contacting experiments the sampling and analysis of these experiments will require significant planning and coordination of effort because of the large number and nature of the samples. Because of the complexity of the system the disinfection stage of these experiments will need to be carried out for extended periods of time until the system reaches a level judged to be reasonable. The efficiency of the disinfection process in the large-scale systems will be very valuable because of the relevance to the situation in real-life systems.

6.2 Dynamic Flushing of Diesel Recommendations

Table 6.2.1 summarizes the flushing recommendations as derived from the traversing fluorescence technique given in Subsection 3.2. Only tests with diesel fuel were performed due to the significant effort involved in calibrating a particular fluid. In general, the measurements suggest that diesel fuel is readily flushed from copper, iron, and PVC pipe substrates. Because the measurement technique is a direct measure of the contaminant thickness, the uncertainty of the measurement implies an uncertainty as to whether or not the diesel fuel has been completely removed from the surface by flushing with water. Subsection 3.2.1 shows the resolution or uncertainty to be on the order of tens of nanometers. It is likely that water flushing does not completely remove diesel fuel from a pipe surface and that a layer less than the sensitivity of the technique (tens of nanometers) exists. By the same token, it is likely that recontamination of the water would be minimal for the same reasons that flushing may not completely remove diesel fuel. Further study is required to quantify the residue of diesel fuel left behind after flushing and the potential for the residue to re-contaminate a cleaned plumbing system.

As it stands, it is recommended to use the computer software presented in Subsection 5.1.4 to estimate the required flushing time for diesel fuel and apply a safety factor of two or more to the calculated flushing time for straight sections. As shown in Subsection 5.2, longer flushing times would likely to be required for obstructed locations such as valves and cavities. Based on simulation results described in Subsection 5.2, flushing time multipliers of 10 to 20 might be required to ensure contaminant removal from locations that are difficult to flush. Verification of the efficacy of the flushing should be confirmed at desired intervals. In addition, hot water and/or cleaning solutions may be desirable.

64

Table 6.2.1 *Results from Dynamic Flow Fluorescence Measurements*	
Contaminant	Diesel fuel
Did it stick	Yes
Did flushing with water remove it	Yes
If not, what is recommended for removal	NA

6.3 Generalization of Measurement Results to Real World Scenarios

The two basic steps in restoring a building plumbing system following a contamination scenario are first to safely purge the system of the contaminated water. Second, flush or treat the system to eliminate any accumulated contaminant. The most effective methods for removing the accumulated contaminants will be a function of the contaminant, plumbing system materials and design. The preferred method, in terms of simplicity and cost is, of course, conventional flushing using water directly from the water distribution system. Other methods may be faster or more effective at cleaning. Three choices that need to be made are as follows:

1. *Which flushing fluid should be used?*
2. *How should the flushing fluid be introduced?*
3. *How should the system be flushed?*

The fluid could be:
- water as is available from the distribution system,
- water that has been treated with additional chlorine or other disinfectant,
- water that has been treated with surfactants or other chemicals to neutralize or react with the contaminant, or
- hot water or steam
- germinant solution to promote spore germination

The fluid source could be:
- from the distribution system,
- from a reservoir supplied by the distribution system, or
- from tanker truck.

The flushing method could be:
- conventional flushing, effuse down drain,
- conventional flushing, effuse collected for disposal,
- flood system and let stand, then drain effuse,
- flood system and let stand, then collect effuse for disposal,
- high velocity pumping,
- steam injection, or
- pulsating flow.

Generally, the most effective method for removing chemical contaminants is by continuous flushing. Accumulated contaminants tend to become entrained in the water due to turbulence, advection, and diffusion. Therefore, removal is primarily a function of the amount of clean water passing through the system. This approach works best for pipe sections, but it is less efficacious for water tanks or reservoirs, especially those that have a top outlet. Flush water velocities are very low in tanks with large diameters relative to pipe diameters, which is usually the case. So, the interaction of flush water with any contaminant that has accumulated on surfaces exposed to the water is slight. Given enough time/water, water soluble contaminants will tend to be removed from the plumbing system. The amount of time/water required to reduce contaminant residuals to a safe level depends upon many factors. These include the type and severity of the contamination, the design of the plumbing system, and the residual levels considered to be safe. It is beyond the scope of this paper. Previous contamination events and measurements with immiscible organic substances suggest that flushing times on the order of days may be required in some cases (Kedzierski, 2006; Morbidity Reports, 1981).

A software tool has been developed to assist in the determination of recommended flush times for contamination events. The details of the software tool are described in Subsection 5.1.3. The general concept is based on modeling the interaction between the contaminant and the pipe walls under different flow conditions to predict the contaminant accumulation and removal. Certain parameters are used as inputs to the software tool. These include mass fraction of contaminant and exposure time, flushing Reynolds number, and the tool provided as output a recommended flushing time. Currently, the software tool has been validated using detailed measurements of diesel fuel contaminant, but is under development to expand its capabilities to cover a wider range of contaminants.

For immiscible contaminants, flushing with hot water may help dissolve the contaminant and shorten required flushing times. Immiscible contaminants that float are best removed from water tanks by flushing out through the top, rather than draining from the bottom. The latter procedure allows a high concentration of the contaminant to come into direct contact with the sediments that tend to accumulate at the bottom of the tank, which makes removal more difficult. In contrast, immiscible contaminants that sink should be drained from the bottom of the tank, and the sediments flushed out if possible.

Bacterial contaminants are best attacked by flooding the plumbing system with disinfectants such as chlorine, and letting it stand to kill or otherwise disable the bacteria. A short flushing will restore clean water (Reipa et al., 2005). For spores, the plumbing system should be flooded with germinant solution, such as an inosine or L-alanine solution, which will encourage the spores to germinate making them susceptible to killing by disinfection in the same manner as bacteria. In both cases, use of hot water enhances the decontamination effectiveness. Also, since the disinfection and growth media solutions only need to flood the system, as opposed to a continuous flushing, only a modest amount of solution is required.

Flushing with water from the water distribution system is the simplest method. It is difficult to control water velocities since they depend on water pressure, which varies with location and function of demand. Sequential flushing of individual water lines will provide the highest water velocities, and, therefore, the greatest scrubbing. Flow control devices, such as faucet aerators and showerheads, should be removed and cleaned individually before flushing. It is important to ensure that the number of water lines being flushed simultaneously does not overwhelm the capacity of the building drainage system, since drainage lines are sized based on the expectation of some probabilistic usage pattern. Additional chlorine or other disinfectant or cleaner could be introduced into the flushing water, either directly from the distribution system or from an auxiliary source closer to the contaminated water system. Care should be taken to ensure that the concentrations of disinfectant or cleaner will not damage the plumbing system materials.

Another concern is a contaminant might become an inhalation hazard due to volatilization into the air after exiting a faucet or other fitting above a sink or tub. Where that is a consideration, avoiding this problem might require special flushing precautions, such as a direct connection between water supply outlets and drain lines. Following flushing, all sinks, tubs and other surfaces that have been exposed to the contaminated water should be thoroughly cleaned.

Water-using appliances present another set of challenges regarding decontamination. The greatest concerns lie with those that involve water that may be consumed or come into contact with building occupants. Chief among these are water tanks, such as hot water heaters, that tend to accumulate sediments and deposits. These are difficult to flush due to their large volumes and corresponding low flow rates. Cleaning water tanks may require direct draining and filling with special cleaning solutions using one of the techniques described below. In some cases, it may not be possible to eliminate all of the accumulated contaminant, and water lines, fittings, fixtures or appliances may need to be replaced. Appliances that have no drain provision, such as residential ice makers, will not be able to be flushed so will need to be removed, cleaned, or replaced. Appliances such as dishwashers and clothes washers cannot be flushed, but they can be cleaned through operation or disconnected to allow their supply lines to be flushed.

Figure 6.3.1 shows a schematic depiction of a possible configuration for flushing a building water supply system. This is done without using water directly from the distribution system by connecting to an external water spigot or other similar connection point. It may be necessary to bypass or disable any back-flow prevention device to allow water to flow in the reverse direction. The flushing water can be pre-treated as is appropriate with disinfectant or cleaner before injection. If the valve at the water meter is closed, the flushing water will be forced under pressure through the water heater and both the cold and hot water supply lines to any fitting or fixture (e.g. faucet, shower, etc.) that is open. In this configuration, the flush water would be directed into the sanitary drains. It could also be collected and transported for disposal. The water supply line leading from the service connection would need to be flushed separately in the normal flow direction.

Figure 6.3.2 shows a similar configuration for decontaminating a water heater tank or for using the water heater as an injection point for flushing the building water supply system. This operation would be similar to that described previously. Although in this configuration, cleaning fluid could be both injected and extracted directly from the water heater via the drain valve. This could be repeated as many times as necessary.

7.0 FLUSHING RECOMMENDATIONS AND CONCLUSIONS

This report presents an overview of measurements and analysis of contaminant accumulation. It includes removal in building plumbing systems and methods for decontaminating building plumbing systems, restoring them to safe operation based on both specific and generic contaminant characteristics. Some of the measurements were used to develop fundamental models to predict maximum contamination levels and required flushing times. These models were, in turn, used to develop computer software that can be used as part of a response to a contamination event.

Measurements consistently showed that most, if not all, of the contaminants did stick to the plumbing material substrates after the initial exposure, while only some (diesel fuel, toluene) showed a substantial reduction from flushing with clean tap water. Others required the addition of high levels of chlorine to effect removal (phorate, gasoline, biologicals).

For the various measurements that were conducted, both exposure and flushing conditions varied considerably. The results were surprisingly consistent regarding the tendency of the particular contaminants to stick and the difficulty of removing them by flushing with tap water. It is not clear if water flushing alone can effectively remove strychnine, phorate or cyanide contamination. Even if it did, it is likely that the required water volumes would be so large they would be impractical. A better approach might be to flush with hot water with high chlorine levels or water with detergent.

Specific recommendations that link decontamination procedures to particular contaminants or groups of contaminants with similar characteristics are given in Subsection 7.1. It is hoped that these recommendation will prove useful as a starting point for a set of comprehensive guidelines that support general response plans for effective recover from water supply system contamination events.

7.1 Inactivation and Flushing of Spores, Bacteria, and Ricin

Lacking specific information about a particular contaminant, the recommended decontamination procedures can be determined according to the type of contaminant based on the following categories listed in table 7.1.

Following is a summary of the recommended list of steps for dealing with a potential contamination scenario involving a building plumbing system following a complaint or other indication of a problem.

68

1. *Collect and analyze water samples to determine if the complaint is associated with the presence of a contaminant. Identify and measure the contaminant concentration;*
2. *Determine the extent of the contamination;*
3. *Isolate the contaminated water piping to prevent propagation to uncontaminated piping;*
4. *Locate the source or point of introduction of the contaminant;*
5. *Determine if the contaminated water can be flushed into the waste water system;*
6. *Assess volatilization potential of contaminant if exposed to atmospheric pressure within a building;*
7. *Determine maximum drainage water flow rate per building to prevent overloading the drainage system;*
8. *Run predictive computer software to estimate contaminant accumulation within the plumbing system;*
9. *Run predictive computer software to determine required flush rate and flush duration;*
10. *Flush with water at the appropriate rate, if considered safe;*
11. *Select appropriate decontamination procedure*
    a. *If water- continues flushing*
    b. *If cleaning agent or shock chlorine- select injection point, flush with solution*
12. *If waste water cannot be discharged into the drainage system*
    a. *Collect waste water*
    b. *Back flush where possible*
13. *Verify effectiveness of decontamination effort*
    a. *Analyze water samples*
    b. *Analyze pipe samples*
14. *Determine if remedial measures are needed to restore plumbing system components*
    a. *Clean/replace faucets, valves, aerators, tanks, hoses*
    b. *Possible surface restoration*
15. *Replace and dispose of any components that could not be decontaminated.*

Table 7.1   *General Decontamination Procedures Based on Contaminant Type*

Contaminant Category	Example	Key Methods
Soluble chemicals	Strychnine, Cyanide	For pipes and tanks- Continuous flushing with water, water buffered with chlorine, or water mixed with cleaner
Immiscible chemicals with specific gravity less than one	Diesel fuel, Gasoline	For pipes- Continuous flushing with water, water buffered with chlorine, or water mixed with cleaner For tanks- Flush through drain valve at bottom of tank or water spigot
Immiscible chemicals with specific gravity greater than one	Phorate	Continuous flushing with water, water buffered with chlorine, or water mixed with cleaner For tanks- Drain through drain valve at bottom of tank, and fill with cleaning solution. Repeat as needed
Sediments or particles	Foreign particles	For pipes- Continuous flushing with water, drain from cleanouts where available For tanks- Drain and flush from bottom
Bacteria	E. coli 0157:H7	For pipes and tanks- Flood system with water and disinfectant and let stand, followed by short flush. Repeat as needed
Spores	Bacillus Anthrasis	For pipes and tanks- Flood system with germinant solution and let stand to allow spores to germinate, followed by short flush
Toxins	Ricin	For pipes and tanks- Continuous flushing with water, water buffered with chlorine, or water mixed with cleaner

8.0 REFERENCES

Alimova, A., Katz, A., Siddique, M., Minko, G., Savage, H. E., Shah, M. K., Rosen, R. B., and Alfano, R. R., 2005, "Native Fluorescence Changes Induced by Bactericidal Agents," IEEE Sensors Journal 5: 704-711.

Alliot, L., Bryant, G., Guth, P.S, 1982, J. Chromat. 232, 440-442.

Almeida, J., Wang, L., Morrow, J. B., and Cole, K. D, 2006, "Requirements for the Development of Bacillus anthracis Spore Reference Materials Used to Test Detection Systems," Journal of Research of NIST 111: 205-217.

Almeida, J. L., Harper, B., and Cole, K. D., 2008, "Bacillus Anthracis Spore Suspensions: Determination of Stability and Comparison of Enumeration Techniques," J. Appl. Microbiol. 104: 1442-1448.

Amadeo, J. P., Rosén C., and Pasby, T. L., 1971, Fluorescence Spectroscopy An Introduction for Biology and Medicine, Marcel Dekker, Inc., New York, p. 153.

American Water Works Association Method 2320, 1991a, "Alkalinity."

American Water Works Association Method 5310 C "Total Organic Carbon by Persulfate-UV or Heated Persulfate Method," 1991b.

American Water Works Association Method 2310, 1990, "Acidity."

ASTM Method D1125-95, 1999, "Standard Test Methods for Electrical Conductivity and Resistivity of Water."

ASTM Method D5790-95, 1995, "Standard Test Method for Measurement of Purgeable Organic Compounds in Water by Capillary Column Gas Chromatography/Mass Spectrometry."

ASTM Method D512-89, 1989, "Standard Test Methods for Chloride Ion in Water". Guilbault, G. G., 1967, Fluorescence: Theory, Instrumentation, and Practice, Edward Arnold LTD., London, 91-95.

Budavari, S., O'Neil, M.J., Smith, A., Heckelman, P., and Kinncary, J.F., 1996, The Merck Index, 12th Edition, Merck and Co., Inc., NJ.

Camper, A. K., Brastrup, K., Sandvig, A., Clement, J., Spencer, C., and Capuzzi, A. J., 2003, "Effect of Distribution System Materials on Bacterial Regrowth," Journal AWWA 95: 107-121.

Cole, K. D., Gaigalas, A. K., and Almieda, J. L., 2008, "Process Monitoring The Inactivation of Ricin and Model Proteins by Disinfectants Using Fluorescence and Biological Activity," Biotech. Prog. 24: 784-791.

Doan, L. G., 2004, "Ricin: Mechanism of Toxicity, Clinical Manifestations, and Vaccine Development. a Review," Journal of Toxicology-Clinical Toxicology 42(2): 201-208.

Endo, Y., Mitsui, K., Motizulki, M., and Tsurgi, K, 1987, "The Mechanism of Ricin and Related Toxic Lectins on The Eukaryotic Ribosomes. The Site and The Characteristics of The Modification of The 28s Ribosomal RNA Caused by The Toxin," J. Biol. Chem. 262: 5908-5912.

EPA, National Primary Drinking Water Regulations, 2005, http://www.epa.gov/safewater/mcl.html#mcls, U.S. Environmental Protection Agency.

EPA Method 7473, 1998, "Mercury in Solids and Solutions by Thermal Decomposition Amalgamation and Atomic Absorption Spectrophotometry."

EPA Method 9213, 1996, "Potentiometric Determination of Cyanide in Aqueous Samples and Distillates with Ion-Selective Electrode."

EPA Method 551.1, 1995a, "Determination of Chlorination Disinfection Byproducts, Chlorinated Solvents, and Halogenated Pesticides/Herbicides in Drinking Water by Liquid-Liquid Extraction and Gas Chromatography with Electron-Capture Detection".

EPA Method 524.2, 1995b, "Measurement of Purgeable Organic Compounds in Water by Capillary Column Gas Chromatography/Mass Spectrometry."

EPA Method 502.1, 1989a, "Volatile Halogenated Organic Compounds in Water by Purge and Trap Gas Chromatography."

EPA Method 503.1, 1989b, "Volatile Aromatic and Unsaturated Organic Compounds in Water by Purge and Trap Gas Chromatography."

EPA Method 180.1, 1978, "Turbidity (Nephelometric)."

Fitzgerald, L. A., Almeida, J. L., and Cole, K. D, 2009, "Enhanced Chlorine Disinfection of Bacillus Thuringiensis Spores by High Flow Rate in a Simulated Drinking Water System," submitted.

Gaigalas, A. K., Cole, K. D., Bykadi, S., Wang, K., and DeRose, P., 2007, "Photophysical Properties of Ricin," Photochemistry and Photobiology 83: 1-8.

Hamid, Z.A., Aal, A.A., 2009, Surface Coatings Tech. 203, 1360-1365.

Hawkins, C. L., Pattison, D. I., Davies, and M. J., 2003, "Hypochlorite-Induced Oxidation of Amino Acids, Peptides and Proteins," Amino Acids 25: 259-274.

Hong, F., Pehkonen, S. J. Agric. Food Chem. 46, 1998, 1192-1199.

Hosni, A. A., Shane, W. T., Szabo, J. G., and Bishop, P. L., 2009, "The Disinfection Efficacy of Chlorine and Chlorine Dioxide As Disinfectants of Bacillus Globigii, a Surrogate For Bacillus Anthracis, in Water Networks," Can. J. Civ. Eng. 36: 732-737.

Ismail, I., Abdel-Monem, N, Fateen, S.E., Abdelazeem, 2009, W. *J. Hazardous Materials*, 166, 978-983.

Kays, W. M., and Crawford, M. E., 1980, Convective Heat and Mass Transfer, McGraw-Hill, New York, NY.

Kedzierski, M. A, 2008, "Diesel Adsorption to PVC and Iron During Contaminated Water Flow and Flushing Tests," NISTIR 7520, U.S. Department of Commerce, Washington, D.C.

Kedzierski, M. A, 2006, "Development of a Fluorescence Based Measurement Technique to Quantify Water Contaminants at Pipe Surfaces During Flow," NISTIR 7355, U.S. Department of Commerce, Washington, D.C.

Kedzierski, M. A., 2002, "Use of Fluorescence to Measure the Lubricant Excess Surface Density During Pool Boiling," Int. J. Refrigeration, 25, 1110-1122.

Ku, Y., Lin, H., 2002, *Water Res* 36, 4155-4159.

Martys, N. S., 2001, "A Classical Kinetic Theory Approach to Lattice Boltzmann Simulation," Int. J. Mod. Phys. C, 12, No. 8, 1169-1178.

Martys, N. S., 1994, "Fractal Growth in Hydrodynamic Dispersion Through Random Porous Media," Phys. Rev. E 50, 335-342.

Matz, L. L., Beaman, T. C., and Gerhardt, P, 2001, "Chemical Composition of Exosporium From Spores of Bacillus Cereus," J. Bacteriol. 101: 196-201.

Mays, Larry, 2000, Water Distribution Systems Handbook, McGraw-Hill, New York.

Miller, J. N., 1981, Volume Two Standards in Fluorescence Spectrometry, Chapman and Hall, London, 44-67.

Morbidity and Mortality Weekly Reports, 1981, "Chlordane Contamination of a Public Water Supply- Pittsburgh, Pennsylvania," Morbid Mortal Wkly Rep 33, 687-693.

Morrow, J. B., Almeida, J. L., Fitzgerald, L. A., Cole, K. D, 2008, "Association and Decontamination of Bacillus Spores in a Simulated Drinking Water System," Water Research, 40: 5011-5021.

Morrow, J. B., and K. D. Cole, 2009, "Enhanced Decontamination of Bacillus Spores in a Simulated Drinking Water System by Germinant Contact," Environ. Eng. Sci. 26: 993-1000.

Nightingale, Z. D., Lancha, A. H., Handelman, S. K., Dolnikowski, G. G., Busse, S. C., Dratz, E. A., Blumberg, J. B., and Handelman, G. J., 2000, "Relative Reactivity of Lysine and Other Peptide-Bound Amino Acids to Oxidation by Hypochlorite," Free Radical Biology and Medicine 29: 425-433.

Radnedge, L., Agron, P. G., Hill, K. K., Jackson, P. J., Ticknor, L. O., Keim, P., and Andersen, G. L., 2003, "Genome differences that distinguish Bacillus anthracis from Bacillus cereus and Bacillus thuringiensis," Appl Environ Microbiol 69: 2755-2764.

Reipa, V., Almeida, J., Cole, K., 2005, "Long-Term Monitoring of Biofilm Growth and Disinfection Using a Quartz Crystal Microbalance and Reflectance Measurements," National Institute of Standards and Technology, Gaithersburg, MD, 20899.

Rice, E. W., Adcock, N. J., Sivaganesan, M., and Rose, L. J., 2005, "Inactivation of Spores of Bacillus Anthracis Sterne, Bacillus Cereus, and Bacillus Thuringiensis Subsp. Israelensis by Chlorination," Applied and Environmental Microbiology 71: 5587-5589.

Rose, L. J., Rice, E. W., Jensen, B., Murga, R., Peterson, A., Donlan, R. M., and Arduino, M. J., 2005, "Chlorine Inactivation of Bacterial Bioterrorism Agents," Applied and Environmental Microbiology 71: 566-568.

Schwartz, T., Hoffmann, S., and Obst, U., 2003, "Formation of Natural Biofilms During Chlorine Dioxide and U.V. Disinfection in a Public Drinking Water Distribution System," Journal of Applied Microbiology 95(3): 591-601.

Szabo, J. G., Rice, E. W., and Bishop, P., 2007, "Persistence and Decontamination of Bacillus Atrophaeus on Corroded Iron in a Model Drinking Water System," Appl Environ Microbiol 73: 2451-2457.

Szabo, J., Rice, E., and Bishop, P., 2006, "Persistence of Klebsiella Pneumoniae on Simulated Biofilm in a Model Drinking Water System," Environmental Science & Technology 40: 4996-5002.

Treado, S.J., 2007, "The Decontamination of Building Plumbing Systems- Analysis and Procedures," NISTIR 7448, National Institute of Standards and Technology, Gaithersburg, MD 20899.

Treado, S., 2005, "Technical Issues Related to the Measurement and Modeling of Contaminant Transport and Accumulation in Building Plumbing Systems," NISTIR 7253, National Institute of Standards and Technology, Gaithersburg, MD, 20899.

Treado, S., Kedzierski, M., Watson, S., Bentz, D., Martys, N., Cole, K., 2006, "Report of Phase one Measurement and Analysis of Building Water System Contamination and Decontamination," <u>NISTIR 7351</u>, National Institute of Standards and Technology, Gaithersburg, MD, 20899.

Wingender J, Flemming HC., 2004, "Contamination potential of drinking water distribution network biofilms," 1: Water Sci Technol,49(11-12):277-86

Appendix A. ChemImage: NIST Pipe Contamination Study

Researcher: John Maier

7301 Penn Avenue, Pittsburgh, PA 15208
Tel 412.241.7335. Fax 412.241.7311.
www.chemimage.com

realize your vision.

REPORT

Title: NIST pipe contamination study
Type of Report: Contract Research

Principal Investigator:

John Maier

maier@chemimage.com

Date of Publication: June 17, 2008

Contracting Officer's Representative:

ATTN: Stephanie Watson

Address:

Building & Fire Research Laboratory 100 Bureau Drive, Stop 8615 Building 226, Room B344 Gaithersburg, MD 20899-8615 **Tel:** 301-975-6448 **E-mail:** Stephanie.Watson@nist.gov

Contractor Name: ChemImage Corporation Contractor Address: 7301 Penn Avenue, Pittsburgh, PA 15208.

Executive Summary

This report focuses on the methods used to evaluate a series of pipe samples exposed to two different contaminants. The work was performed for Stephanie Watson on NIST in May of 2008. In this project fluorescence spectral imaging with UV illumination in conjunction with PCA image analysis was used to study the contamination of different types of pipe material with different contaminants. The limitations of this approach include no sensitivity to contamination that does not change the luminescence of the sample under UV light illumination and the lack of specific chemical information about the chemical nature of the residues left on the pipes.

With those limitations in mind, this analysis shows that the rubber pipe material does not suffer significant residual contamination for either contaminant. Brass and PVC similarly do not show any evidence of residual contamination with mercuric chloride. Brass shows evidence of complete residual contamination with strychnine while the PVC has 78 % surface area contaminated with strychnine. The iron samples show evidence of complete residual contamination with both the mercuric chloride and strychnine contaminants used in this study. The copper samples show residual contamination for both contaminants, but not complete contamination in either case. For the case of strychnine 66% of the surface appears to have residual contamination while for the case of mercuric chloride only 9% of the surface appears to have residual contamination.

Background

NIST is involved in a research project to study the interaction of two fouling contaminants on 5 different types of pipe. The basic question is: after an extended exposure of the contaminants and a washing step with water, how much of the surface remained contaminated. After discussion with ChemImage, an experimental approach was designed including a preliminary set of experiments to assess what spectral imaging methods might help answer this question and a second set of more focused experiments chosen based on performance in the preliminary work.

Materials and Methods

Samples (provided by NIST):
- • 5 unexposed pipe samples
  - o Copper
  - o Brass
  - o Iron
  - o Rubber
  - o PVC
- 5 pipe samples exposed to mercuric chloride (HgCl$_2$)
- 5 pipe samples exposed to Strychnine
  - • Samples of chemicals exposed to pipe and precipitants from pipes
    - o Strychnine
    - o HgCl$_2$
    - o Precipitant from iron exposed to mercuric chloride
    - o Precipitant from brass exposed to mercuric chloride
    - o Precipitant from copper exposed to mercuric chloride

Preliminary work

(No unexposed samples were available for this phase of work)

Raman spectroscopy and Imaging

Raman spectroscopy was performed using ChemImage's Falcon II Raman microimaging system. The spectra were acquired using 532 nm excitation. Data collection parameters varied between samples. Raman spectra were collected for the pure components and the precipitant samples. The pure component materials, mercuric chloride and strychnine, produced very strong Raman scattering with a high signal to noise (SNR) ratio; however the precipitant samples were highly fluorescent and the Raman bands were obscured by the abundant fluorescence. Figure 1 shows a Raman spectral library of the pure component and precipitant samples provided by NIST.

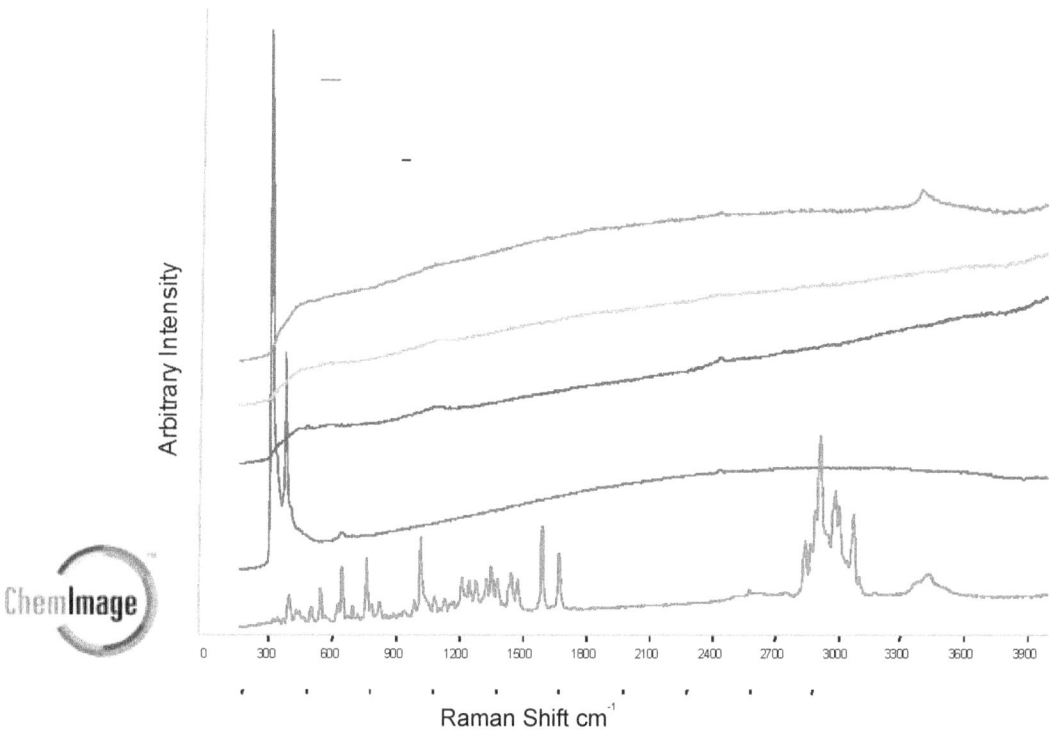

Figure 1: Raman spectra of pure components and precipitant samples.

The spectra shown in Figure 1 have not been baseline corrected, but have been normalized using vector normalization. They are displayed on the same y-axis with an offset to facilitate visual comparison. The Raman peaks are sharp and distinct while the luminescence is a broad background spectral response.

Fluorescence and IR macro imaging

In the preliminary phase of this study, the exposed pipe samples were analyzed using Macroscopic Near-Infrared Chemical Imaging, Visible Absorbance Chemical Imaging and Fluorescence Chemical Imaging. The five pipe samples exposed to mercuric chloride and the five pipe samples exposed to strychnine (a total of 10 samples), were placed in a single field of view and a dataset was collected for each mode. In addition to the sample data collection, a background image was collected (for each mode) using the same respective parameters.

The parameters used for each data collection are as follows:

Near Infrared Imaging Spectral Range: 850 – 1800 nm Step Size: 5 nm Exposure Time: 16.36 msec Averaging (per frame): 10 Binning: None Processing: Divide by background image, Convert to Absorbance, Extract wavelengths 950 – 1650 nm, Normalize

Visible Reflectance Imaging Spectral Range: 420 – 720 nm Step Size: 5 nm

Exposure Time: 0.5 sec Averaging (per frame): 1 Binning: None Processing: Divide by background image, Normalize

Fluorescence Imaging Spectral Range: 420 – 720 nm Step Size: 10 nm Exposure Time: 60 sec Averaging (per frame): 1 Binning: None Processing: Divide by background image, Normalize

Only the fluorescence chemical image of the 10 exposed pipe samples yielded a promising result. In order to better understand the information contained in the fluorescence chemical image, it was necessary to analyze unexposed pipe samples along with the exposed samples.

Phase 2 work

Phase 2 work was started after discussing the preliminary results with Stephanie Watson of NIST. For phase 2 of the project, the control (un-exposed) pipe samples were placed on the same platform as the exposed pipe samples (15 samples total in the field of view). A fluorescence chemical image dataset was collected for the samples according to the following parameters:

Spectral Range: 420 – 720 nm Step Size: 10 nm Exposure Time: 60 sec Averaging (per frame): 1 Binning: 2 x 2

Analysis

1. Goals

Discussion with Stephanie Watson of the preliminary results led to a second phase of work focused on using spectral fluorescence imaging to get data which would be evaluated to estimate the percent area of the surface that underwent change based on spectral differences. Principal Component Analysis (PCA) was chosen as an established tool for analysis of the spectral images. As described below, percent area changed was calculated based on first measuring the area of each sample in the image and second using PCA images of the samples of a given type of pipe to estimate the area of each sample which is spectroscopically different from normal.

One limitation of this approach is the absence of the specific chemical information about what compounds were causing the spectral differences in the contaminated samples. A second limitation is that a change would not be detected using this approach unless the interaction of the contaminant and the sample manifested itself as a change in the sample luminescence.

2. Estimation of total area of samples

The area of each sample was estimated using a frame from the visible reflectance image and a two step masking process. A frame was selected in which all of the samples were visible with positive or negative contrast compared to the background. Figure 2 shows a single frame from the visible reflectance image of the samples. A Region Of Interest (ROI) selection tool in ChemImage Xpert was used to select a region containing samples of similar intensity and contrast relative to the background. A first binary mask was created in the shape of the selected region of interest and multiplied by the image. This results in an image of the same size as the original image, but only containing information within the boundaries of the selected region. A histogram of this image shows two distinct non-zero populations, one for the background and one for the sample. A second mask is created using a histogram based threshold tool in ChemImage Xpert. In this second mask a pixel is set to one if the pixel is in the population which corresponds to the samples and zero if it is not.

This process is carried out so that ultimately there is a mask for each sample. The mask is used subsequently for two parts of the process. First, the sum of the intensity in a region of the mask containing a sample is a value proportional to the area of the sample. Second, the mask is also used to multiply the spectral fluorescence image yielding a fluorescence image which only contains data at

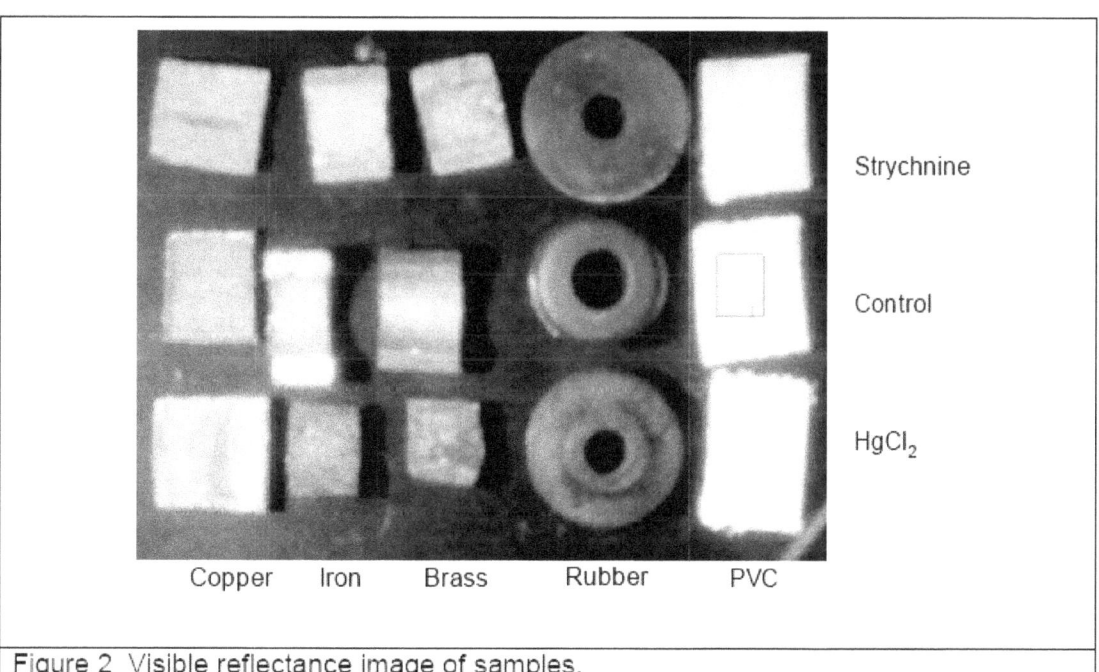

Figure 2 Visible reflectance image of samples.

the site of the specific sample. In this fashion the fluorescence spectral image of only the three copper samples (for example) can be analyzed together, yet separate from the other types of pipe samples. Figure 3 shows the same image as in Figure 2 with a red overlay of the mask created using the above approach. This image demonstrates how accurately the method described above captures the area of the samples.

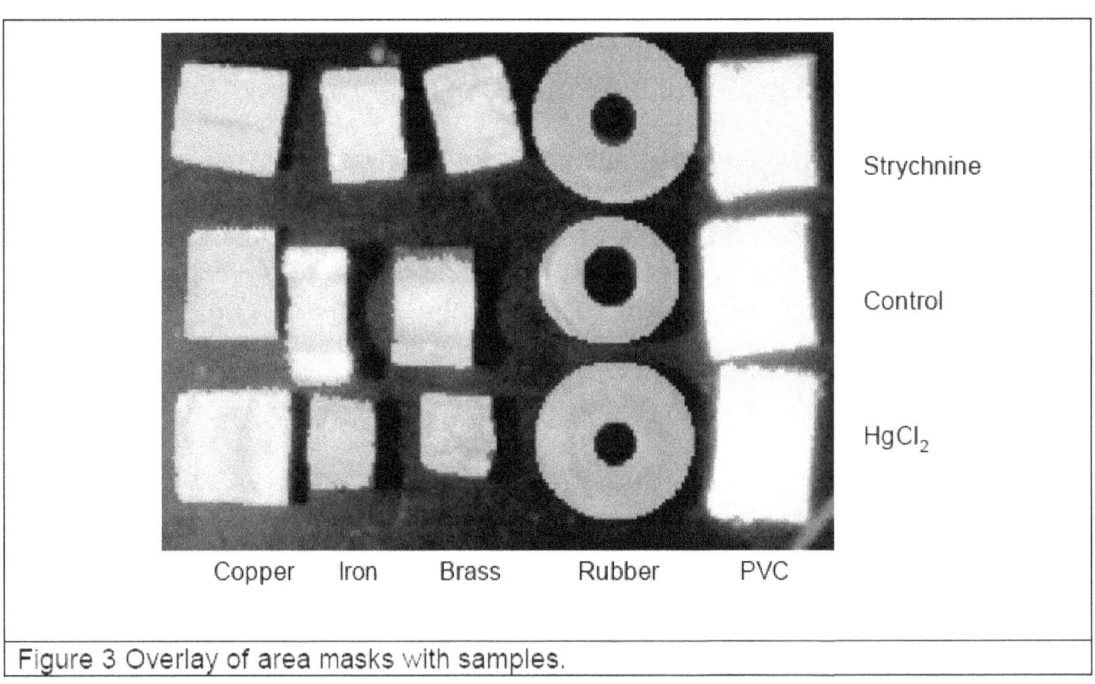

Figure 3 Overlay of area masks with samples.

3. Estimation of affected area of samples

In order to estimate the contaminated area of a set of samples the fluorescence spectral image is masked to contain only information from the set of samples. A mask for the two contaminated samples and the uncontaminated sample of one type of pipe is selected from the mask image of all the samples. This mask image is multiplied by the fluorescence spectral image of all the samples yielding an image which contains information only where the three samples of a given material are present. For the pixels which are not on one of the samples the data values in the image are zero for all spectral frames.

ChemImage Xpert was used to perform PCA on such images of each set of pipe material. The result of the PCA is a set of principal component loadings and a set of score images. The loadings are the orthogonal basis vectors, which result from the principal component analysis. The score images are maps of the weight of each of the principal components, or loadings, at each pixel. PCA is an objective analysis of the mathematical variability in a set of spectra. In the case of this application of PCA, the set of spectra are the spectra from each pixel in the image of a set of samples of one pipe type. Without knowing the chemical origin of changes which cause the spectra to be different, PCA image analysis allows objective assessment of whether spectra from pixels in either exposed sample are similar to, or different from, the unexposed sample. This manifests itself in the score image.

For instance, in a set of spectrally identical samples, the score image would be expected to be have the samples be bright and dark in the same frames. There is a frame of the score image for each principal component of the analysis. Alternatively, two samples of one type and a third sample of another type would lead to score images which would consistently show the like samples having like contrast.

The principal component spectra, or loadings, are mathematically determined, and are not necessarily determined in such a way that specific identifiable spectral signatures are present in the loadings. In fact, if a distinctive spectral signature is present in the sample set, it may mathematically fall as a linear superposition of one or more of the loadings determined through PCA. Various approaches exist to rotate the loadings in a fashion that maintains the orthogonality of the basis set, while aligning the basis vectors with more interpretable spectra. This analysis was not performed in this project because there was not a specific goal of establishing the spectra of different components.

The score images can be evaluated using the same histogram based threshold approach used to determine the masks for estimating the area of each sample. A difference in intensity in a given score image indicates a difference in the material, especially when it is correlated by spatially similar observations in other score images. Evaluation of the histogram of a score image shows graphically when there is obviously more than one component present. Setting bounds on the intensity values for inclusion in a mask will yield a binary mask with ones where pixels with a particular spectral type appear.

4. Example of application of contaminated area estimation approach:

The subject of this study is how much of the surface is contaminated, not just

whether there is a difference overall. Spatial variation of the spectral change which indicates interaction with the contaminant can be analyzed by considering the individual spectra of each pixel, and the degree to which they are different from the spectrum of the normal sample.

An example of the analysis described above and associated data is presented below.

In the first step the mask which selects for the copper samples is multiplied by the spectral fluorescence image yielding the image shown in Figure 4.

Figure 4 is the frame at 560 nm of the masked fluorescence spectral image. For reference, the unexposed sample is in the middle, the strychnine exposed sample is on the top, and the $HgCl_2$ exposed sample is on the bottom (see Figure 2). The gray scale is chosen to span the range of non-zero data in the frame at 560 nm. All of the pixels in this frame which are black have a value of zero, and

Figure 4 frame from spectral image data for copper samples.

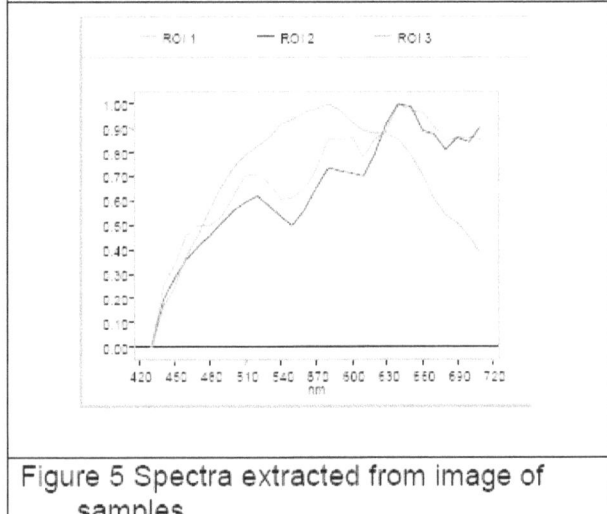

Figure 5 Spectra extracted from image of samples.

will therefore not contribute to the subsequent analysis of spectral variability.

Figure 5 shows the mean spectra of the ROIs shown in Figure 4. Note that ROI3 (the strychnine sample) has a different spectrum from the $HgCl_2$ and unexposed sample indicating the difference of this sample from normal. We assume this difference is due to the exposure. It is clear that the mean spectrum of the strychnine exposed sample is different from normal.

ChemImage A principal component analysis is performed using ChemImage Xpert chemical imaging software. The first frame of the resulting score image is shown in Figure 6 (left). Figure 6 (right) shows the histogram of intensity values for this frame.

This histogram plots the number of pixels (y-axis) as a function of the intensity value (x-axis) where the intensity value is the weight that results from the principal component analysis for the first principal component. This weight is the relative weight of the spectral loadings which also result from PCA. The loadings are the orthogonal basis set which are chosen to account for the variability in the set of spectra (in this case the spectra from the pixels where there is sample material).

The weights in the first principal component for the spectra in this case are negative numbers. This is because the PCA performed here is not run with constraints that the loadings, or principal components, appear like physical spectra (have only positive values). Because the spectra can have what appear as "negative going peaks", the weights are sometimes negative numbers. While this is not relevant for the present analysis, the reason for the negative weights in the histogram can be a point of confusion.

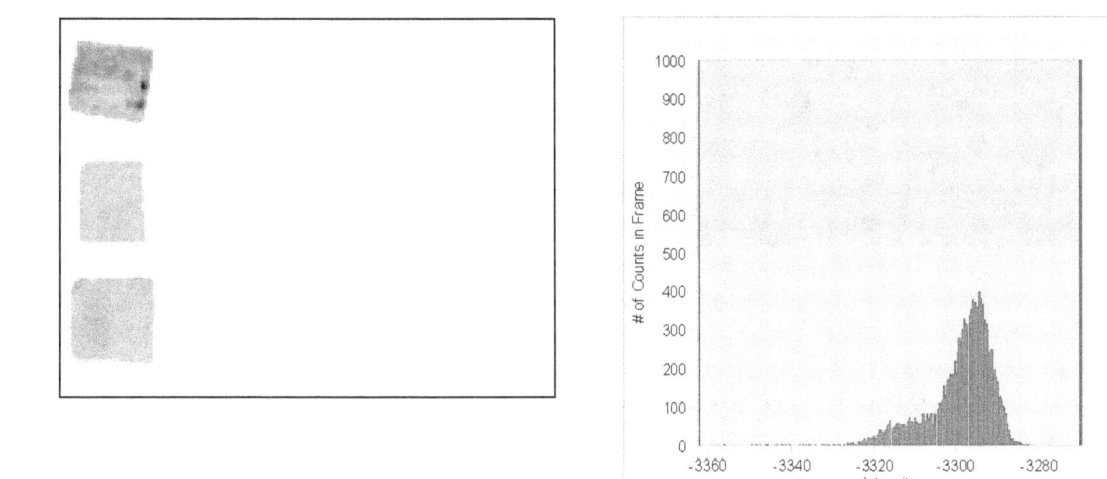

Figure 6 Histogram of score image for PC1 (copper)

The image shows that while the normal sample is homogeneously gray (having similar weights), the strychnine exposed sample is not homogeneous. Similarly, the HgCl2 exposed sample has some dark areas and is also distinct from the normal sample.

The histogram is consistent with two populations of for the weights, one with a peak at about –3295 and one with a peak at about -3310. The histogram can be used to select a population of pixels in the image, by choosing a range of values for construction of a binary mask. In this case, two masks, each encompassing predominantly one distribution, can be generated.

Figures 7 and 8 show masks (left) generated from the upper and lower distributions. The white areas in the mask image are the pixels which have weights in the histogram highlighted by gray. In Figure 7 the lower values (less than –3305) are highlighted in the histogram. The resulting mask has white pixels in the strychnine sample in the spatial distribution which appears like the dark areas seen in the score image shown in Figure 6. The areas are dark in Figure 6 because they have lower weights from the PCA evaluation. The histogram analysis allows identification of those regions based on the observation that there are two distributions in the histogram. By setting the threshold based on observed distributions in the histogram, an image can be generated with less bias than if the image itself is employed by the user.

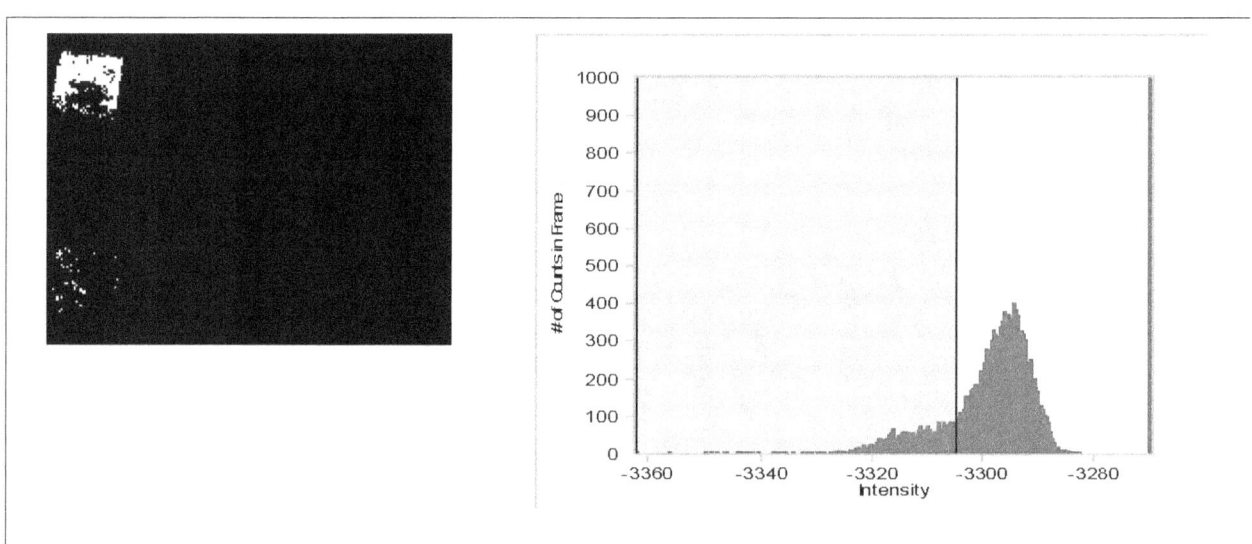

Figure 7 Mask 1 for based on PC1 score image from copper.

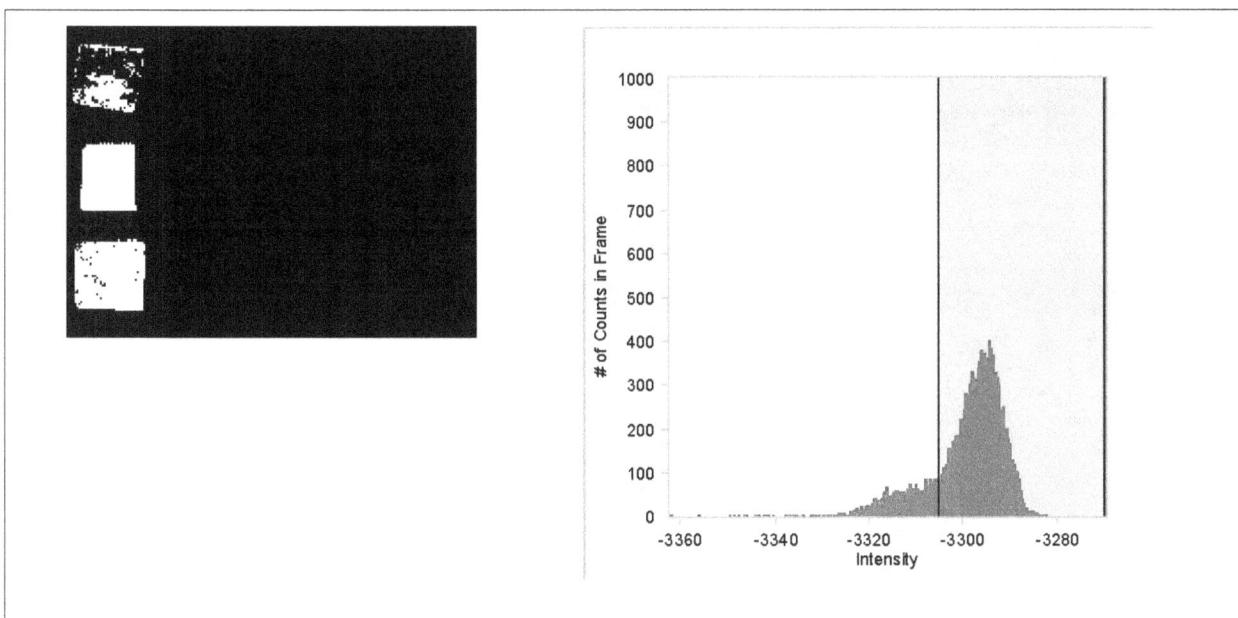

Figure 8 Mask 2 based on PC1 score image of copper.

The resulting mask images can be evaluated in a fashion similar to the original mask images which were used to estimate the sample area for each sample. In this case the number of white pixels in the mask of a specific characteristic (same as control or different from control for example) are added up to determine, in pixels, the interacted area. For instance the upper sample in this example has a significant number of pixels which are not like the normal. These pixels appear white in Figure 7 (note in this mask the normal sample is black). The total number of white pixels in the top sample is 2966. From the analysis of sample area described above the total area of the sample is 4504 pixels. Thus, the contaminated area (the area which is not like normal) is 66%.

In all cases all of the score images were reviewed. In some cases there was evidence of change in one type of pipe sample from normal in one score image, and another exposure in a second score image. This is to be expected as there could be different spectral responses to the exposures for the two different exposing chemicals for a given type of pipe sample.

In this fashion the combination of spectral imaging of samples and controls along with masks generated from histogram analysis of PCA score images can give an estimate of the percent interaction of the pipe samples with the chemicals they were exposed to.

Results / Conclusion

As described above, each set of samples of a particular type were evaluated. The table shown in Figure 9 shows the estimated total area and interacted area of the samples.

Sample	Total Area	affected area	% affected
Strychnine			
Copper	4504	2966	66%
Iron	1924	1924	100%
Brass	3559	3567	100%
Rubber	8154	0	0%
PVC	5199	4057	78%
Untreated			
Copper	4064		
Iron	3568		
Brass	3553		
Rubber	4674		
PVC	5732		
Mercuric Chloride			
Copper	5311	460	9%
Iron	2435	2434	100%
Brass	2241	4	0%
Rubber	7541	0	0%
PVC	5928	0	0%

Figure 9 Results table.

In conclusion, fluorescence spectral imaging with UV illumination in conjunction with PCA image analysis was used to study the contamination of different types of pipe material with different contaminants. The limitations of this approach include no sensitivity to contamination that does not change the luminescence of the sample under UV light illumination and the lack of specific chemical information about the chemical nature of the residues left on the pipes.

With those limitations in mind, this analysis shows that the rubber pipe material does not suffer significant residual contamination for either contaminant. Brass and PVC similarly do not show any evidence of residual contamination with mercuric chloride. Brass shows evidence of complete residual contamination

with strychnine while the PVC has 78% surface area contaminated with strychnine. The iron samples show evidence of complete residual contamination with both the mercuric chloride and strychnine contaminants used in this study. The copper samples show residual contamination for both contaminants, but not complete contamination in either case. For the case of strychnine 66% of the surface appears to have residual contamination while for the case of mercuric chloride only 9% of the surface appears to have residual contamination.

Appendix B. Determination of Soluble Copper Cyanide Complexes in Tap Water

Tuesday, July 8, 2008
Ivy Grimm
William R. LaCourse, Ph.D
University of Maryland Baltimore County
Department of Chemistry and Biochemistry
1000 Hilltop Circle
Baltimore MD, 21250

Purpose:
To identify soluble metal cyanide complexes that form in tap water treated with copper pipe and a cyanide salt, and to determine the concentration of copper cyanide complexes using ion chromatography (IC) with ultra-violet detection (UV).

Experimental:
Method: CuCN2.met
Run Time: 35 min.
Gradient Elution:
  Eluent A: 20 mM NaOH/150 mM NaCN
  Eluent B: 20 mM NaOH/300 mM $NaClO_4$
  Eluent C: 20 mM NaOH
Pump Program:

Time (min)	%A	%B	%C
Initial	10	10	80
0.0	10	10	80
18.0	10	45	45
22.0	10	45	45
25.0	10	10	80
35.0	10	10	80

Flow Rate: 1 mL/min
Detection: UV absorbance at 215 nm
Sample treatment: Filtered through 0.45 μm LC Acrodisc® syringe filter (Waters)
Blank: Tap water from NIST lab
Method Reference: Application Update 147, Dionex Corporation, available at www.dionex.com

Results:
Results obtained from the IC-UV method provided quantitative determination of concentration levels of the copper-cyanide complex, $[Cu(CN)_3]^{2-}$, present in samples. The following data and results were determined using a $[Cu(CN)_3]^{2-}$ standard calibration as a reference. These standards were prepared following the procedure in Application Update 147 (Dionex Corporation). Interpretation of the results is not included.
  Analytical Figures of Merit:

Fig. B.1: Shows the calibration curve for the [Cu(CN)$_3$]$^{2-}$ standard

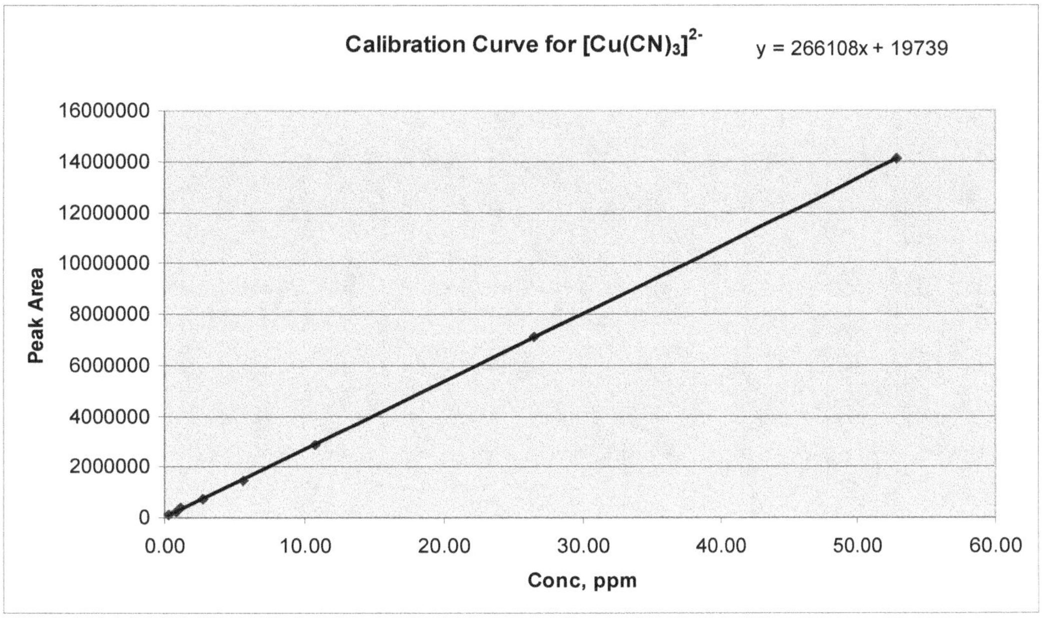

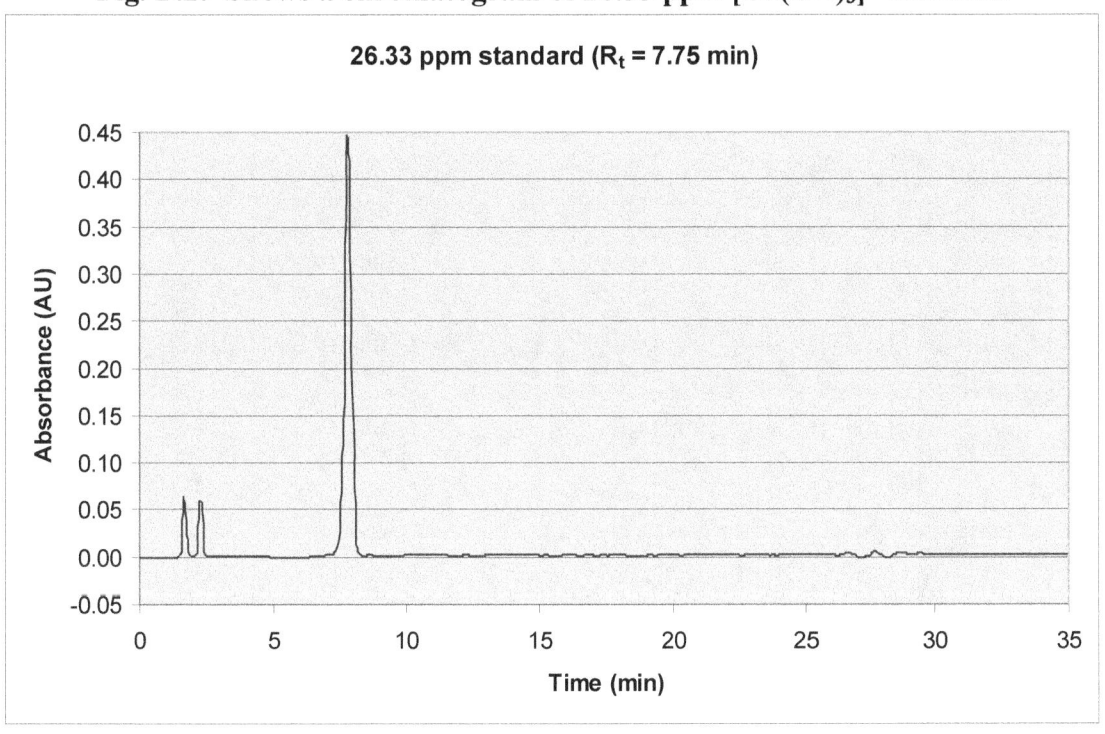

Calibration Curve for [Cu(CN)$_3$]$^{2-}$ y = 266108x + 19739

Sensitivity = 266108, R^2 = 0.9999, LOD = 0.108 ppt

Chromatograms:
Fig. B.2: Shows a chromatogram of 26.33 ppm [Cu(CN)$_3$]$^{2-}$ standard

26.33 ppm standard (R_t = 7.75 min)

Fig. B.3: Shows a chromatogram of sample #4 – 50 ppm KCN + Cu pipe (old)
Mean Concentration = 0.963 ppm, (Zoomed in to enhance peak)

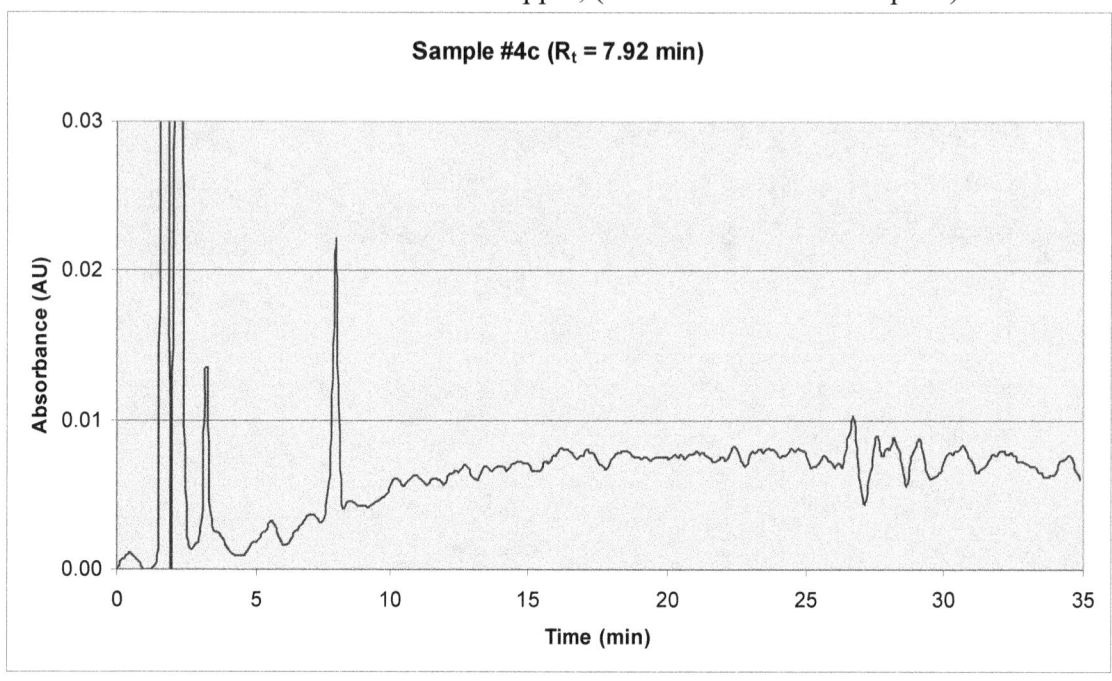

Fig. B.4: Shows a chromatogram of Sample #11 – 50 ppm KCN + Cu pipe (fresh)
Mean Concentration = 59.49 ppm

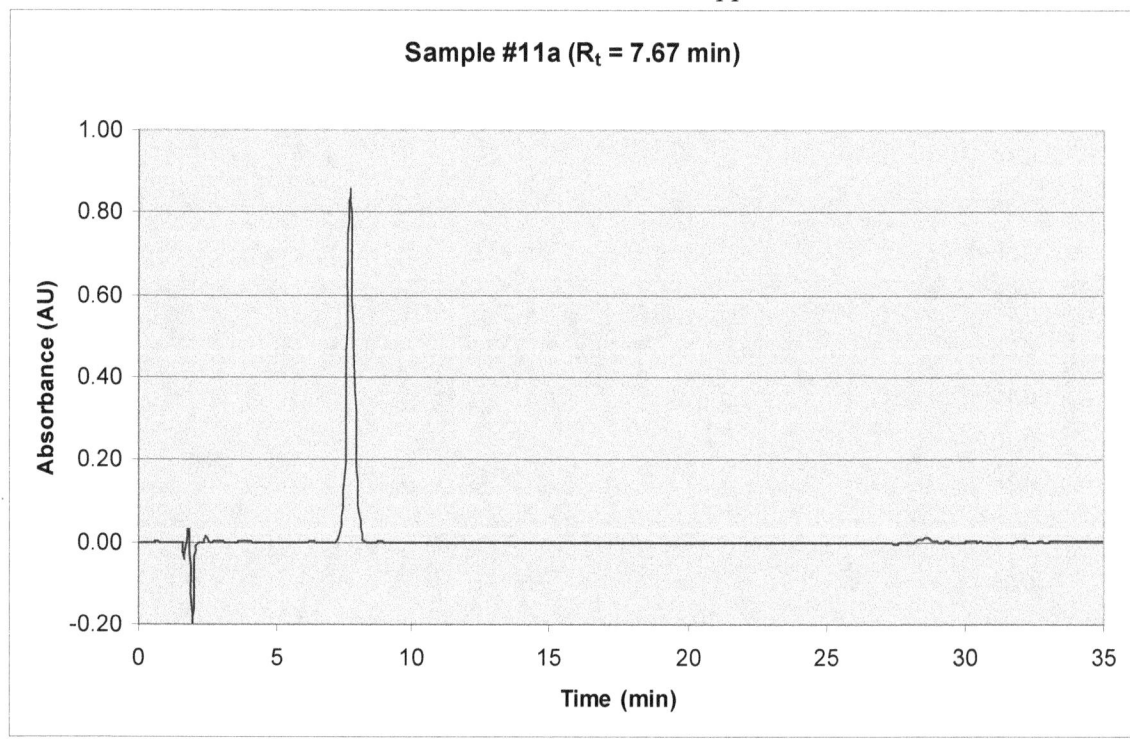

Sample	Conc. $[Cu(CN)_3]^{2-}$ (ppm)	Mean Conc. (ppm)	% RSD (N = 3)
#1 - 3 ppm KCN + Cu (a)	ND		
#1 - 3 ppm KCN + Cu (b)	ND		
#1 - 3 ppm KCN + Cu (c)	ND		
#2 - 10 ppm KCN + Cu (a)	ND		
#2 - 10 ppm KCN + Cu (b)	ND		
#2 - 10 ppm KCN + Cu (c)	ND		
#3 - 20 ppm KCN + Cu (a)	12.53		
#3 - 20 ppm KCN + Cu (b)	12.57	12.55	0.17
#3 - 20 ppm KCN + Cu (c)	12.55		
#4 - 50 ppm KCN + Cu (a)	0.88		
#4 - 50 ppm KCN + Cu (b)	1.12	0.963	14.01
#4 - 50 ppm KCN + Cu (c)	0.89		
#5 - 3 ppm NaCN + Cu (a)	ND		
#5 - 3 ppm NaCN + Cu (b)	ND		
#5 - 3 ppm NaCN + Cu (c)	ND		
#6 - 10 ppm NaCN + Cu (a)	1.04		
#6 - 10 ppm NaCN + Cu (b)	1.01	0.956	12.47
#6 - 10 ppm NaCN + Cu (c)	0.82		
#7 - 20 ppm NaCN + Cu (a)	ND		
#7 - 20 ppm NaCN + Cu (b)	ND		
#7 - 20 ppm NaCN + Cu (c)	ND		
#8 - 50 ppm NaCN + Cu (a)	5.54		
#8 - 50 ppm NaCN + Cu (b)	6.97	5.95	14.82
#8 - 50 ppm NaCN + Cu (c)	5.36		
#9 - 20 g KCN + 1g Cu [†]	ND		
#10 - 10 ppm KCN + Cu fresh (a)	18.76		
#10 - 10 ppm KCN + Cu fresh (b)	18.99	18.76	1.26
#10 - 10 ppm KCN + Cu fresh (c)	18.52		
#11 - 50 ppm KCN + Cu fresh (a)	59.52		
#11 - 50 ppm KCN + Cu fresh (b)	59.57	59.49	0.18
#11 - 50 ppm KCN + Cu fresh (c)	59.37		
3 ppm KCN + tap water (in duplicate)	ND		
10 ppm KCN + tap water (in duplicate)	ND		
50 ppm KCN + tap water (in duplicate)	ND		

† This sample had a broad interfering peak that overloaded the detector. Diluting the sample to overcome this interference would cause the analyte peak to become not detectable.

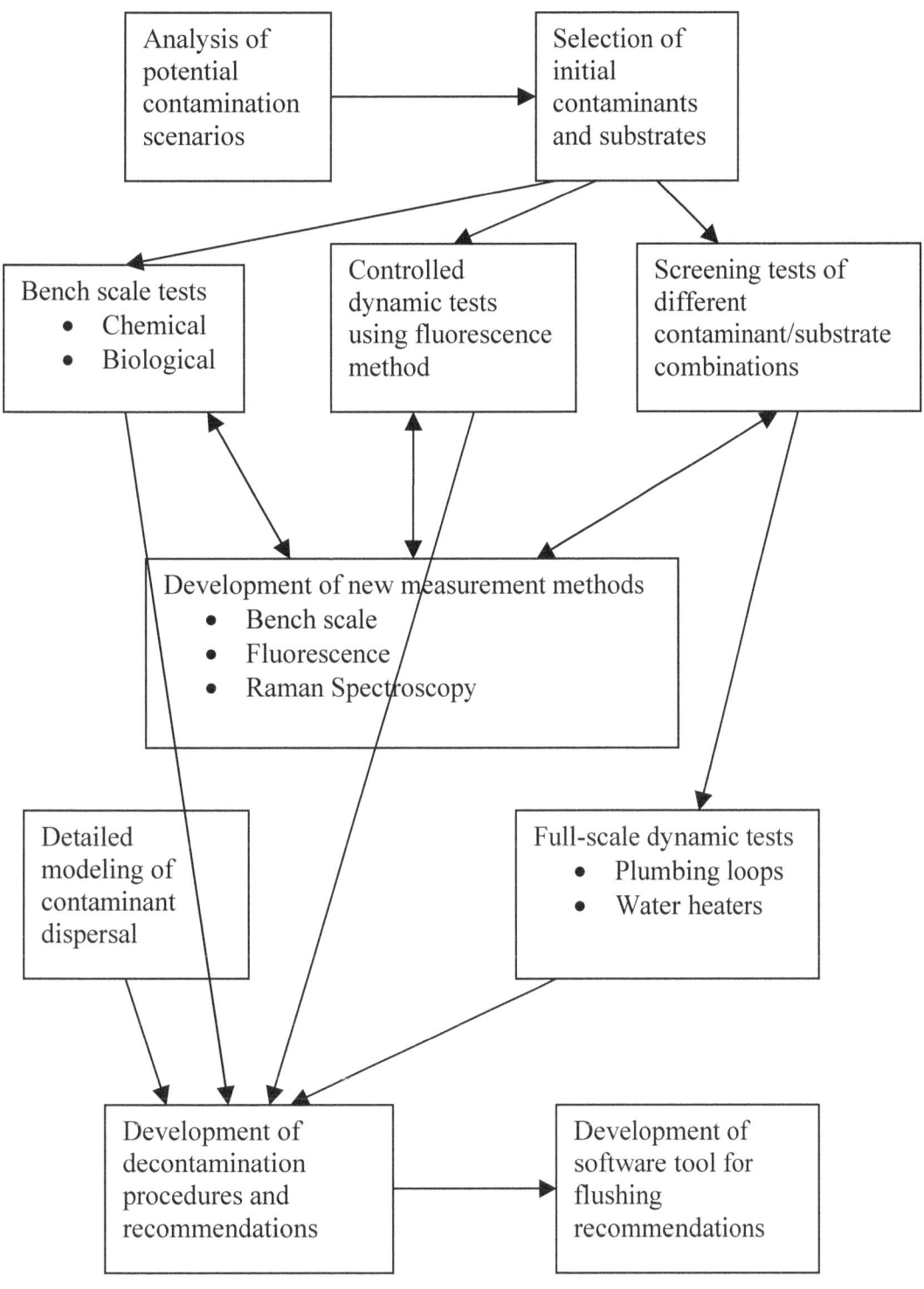

Fig. 1.1 Flowchart of the project and report

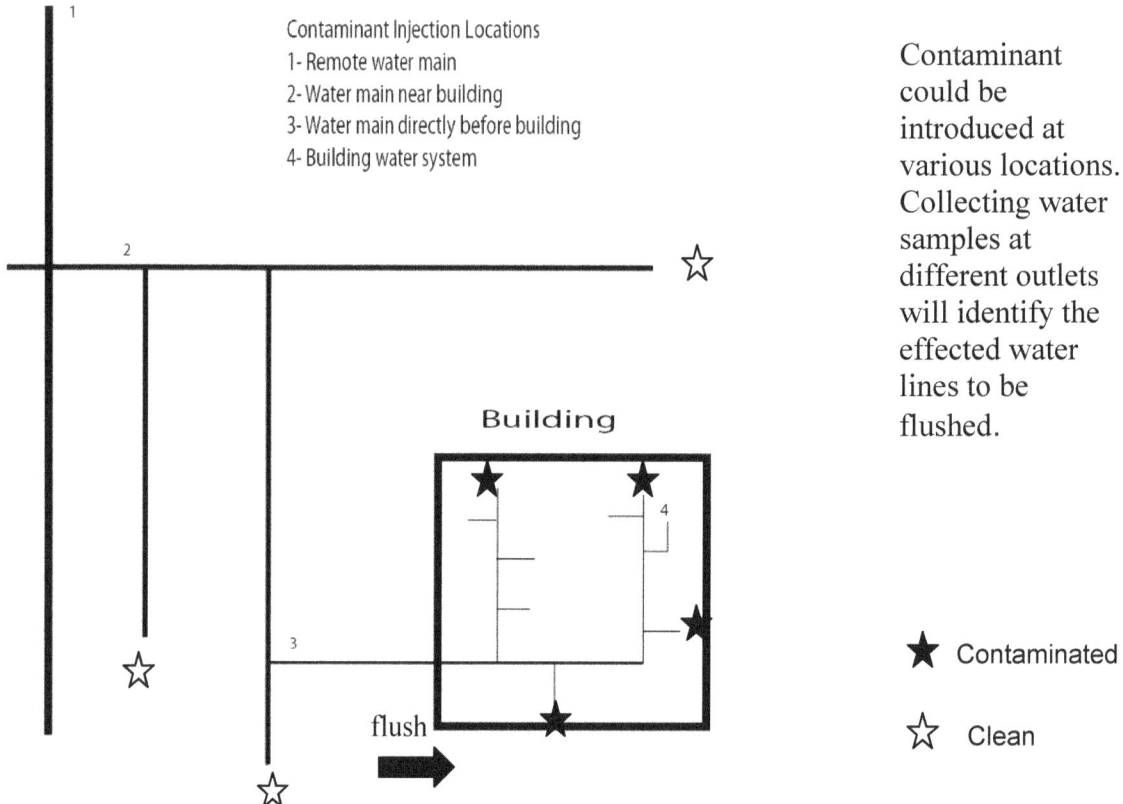

Contaminant Injection Locations
1- Remote water main
2- Water main near building
3- Water main directly before building
4- Building water system

Contaminant could be introduced at various locations. Collecting water samples at different outlets will identify the effected water lines to be flushed.

★ Contaminated

☆ Clean

Building

flush

Fig. 1.2 Collecting and analyzing water samples from various locations will enable effected portions of the plumbing system to be identified

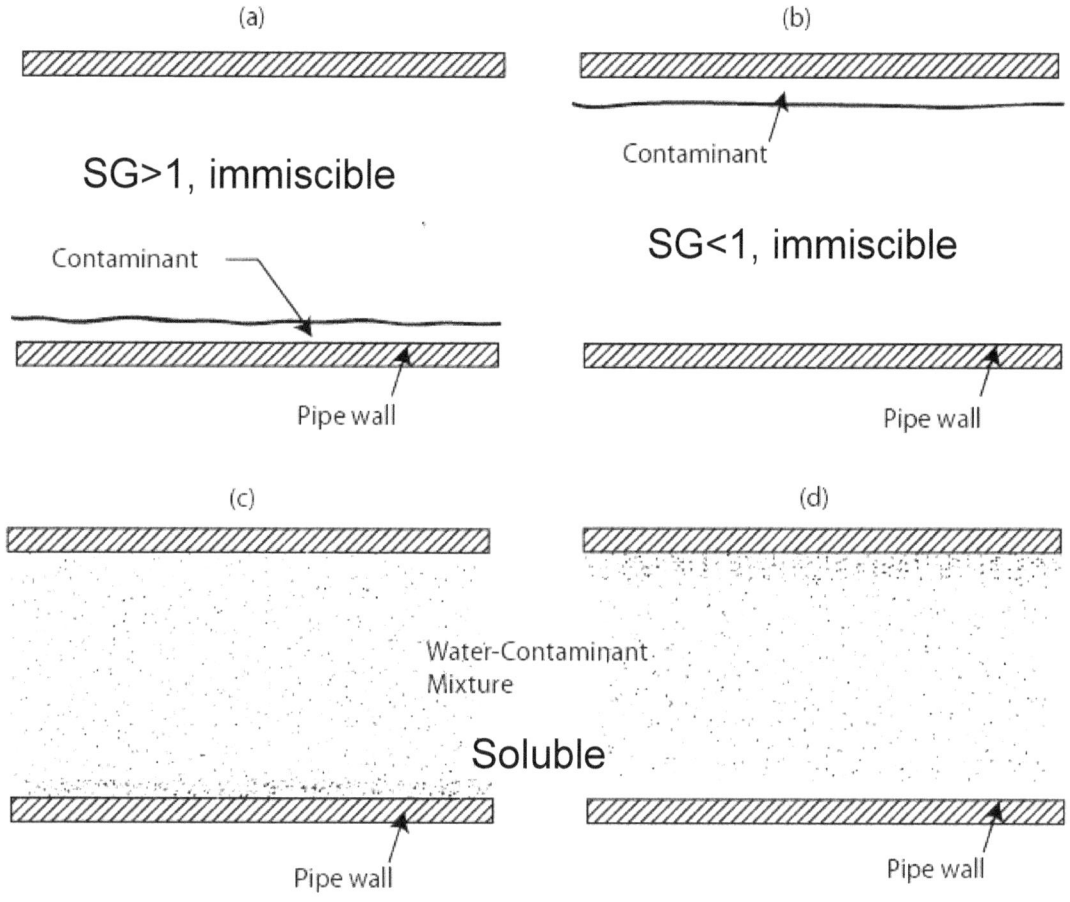

(a)

SG>1, immiscible

Contaminant

Pipe wall

(b)

Contaminant

SG<1, immiscible

Pipe wall

(c)

Water-Contaminant
Mixture

Soluble

Pipe wall

(d)

Pipe wall

Contaminant/substrate exposure can vary with solubility and density

Fig. 1.3 The contact between the contaminant and plumbing system components will depend in part on the contaminant properties.

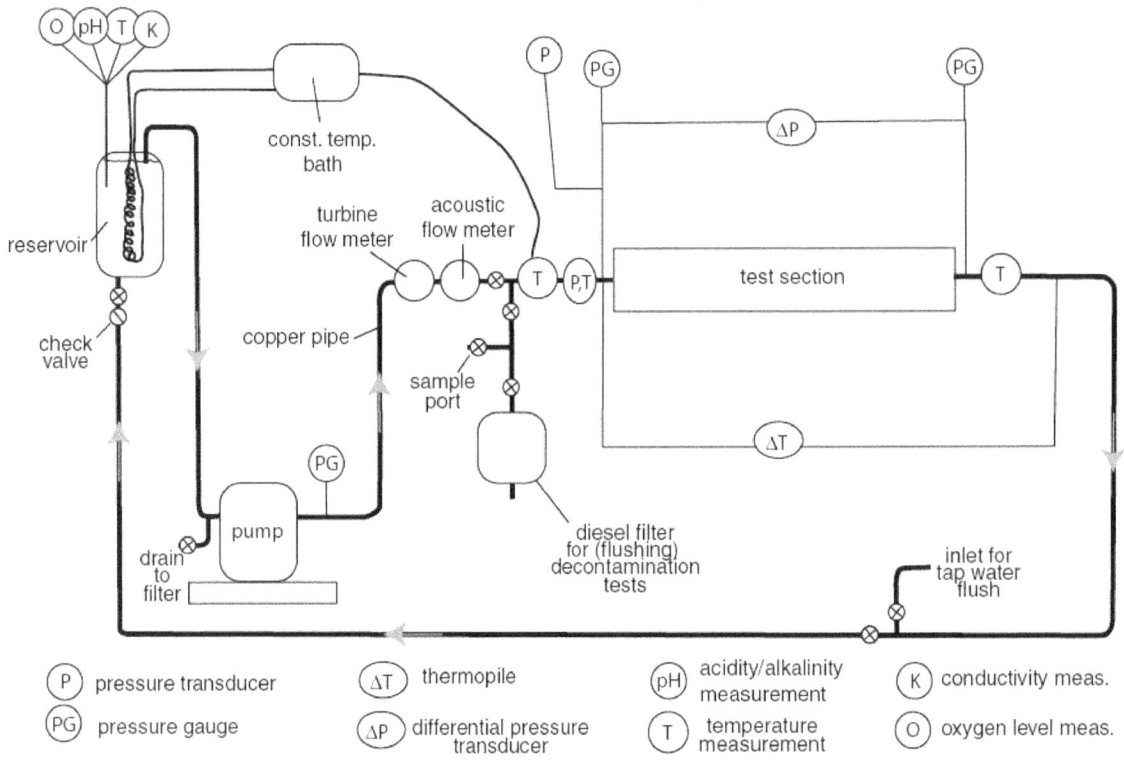

Fig. 3.2.1.1 Schematic of dynamic flow test loop

98

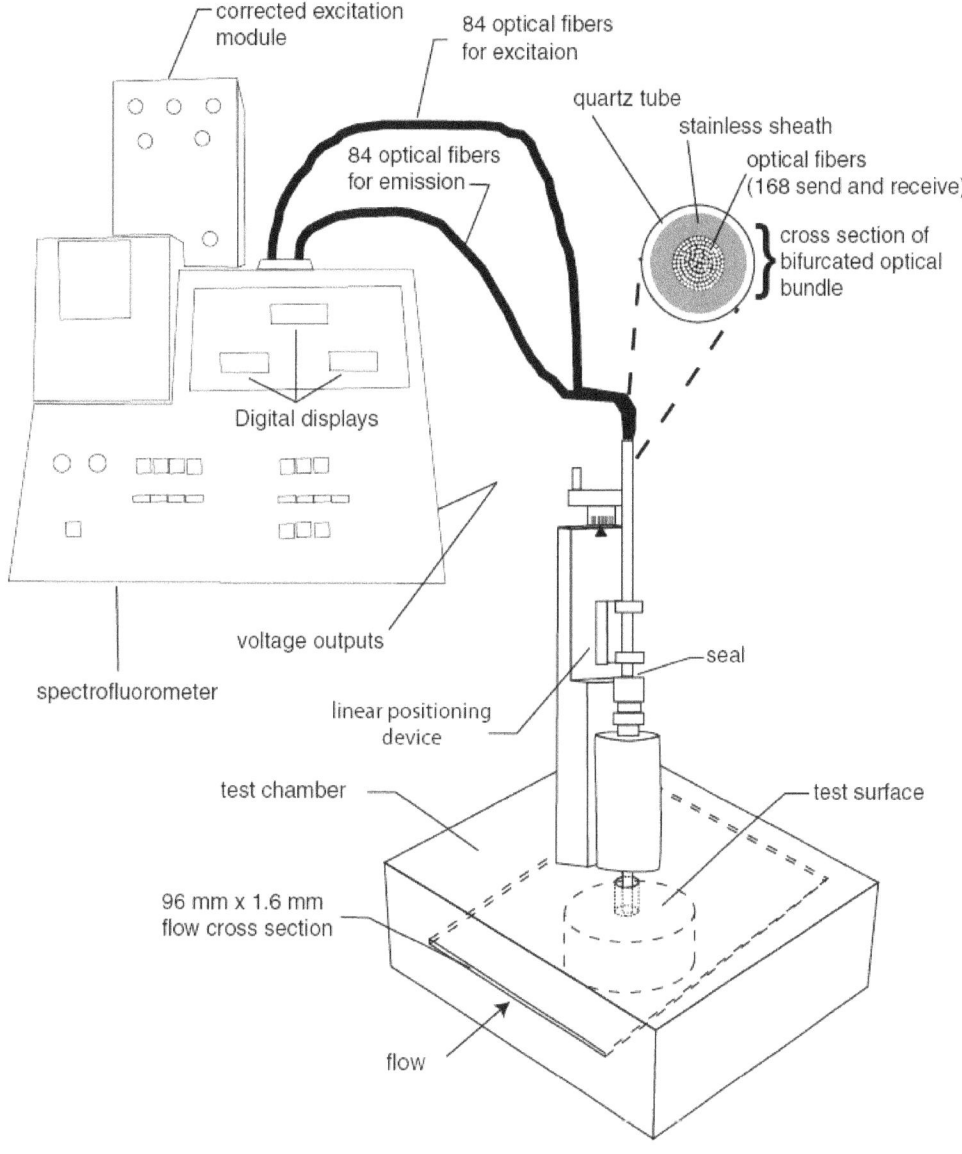

corrected excitation
module

84 optical fibers
for excitaion

quartz tube

stainless sheath

optical fibers
(168 send and receive)

84 optical fibers
for emission

cross section of
bifurcated optical
bundle

Digital displays

voltage outputs

seal

spectrofluorometer

linear positioning
device

test chamber

test surface

96 mm x 1.6 mm
flow cross section

flow

**Fig. 3.2.1.2 Schematic of spectrofluorometer, test section, and linear
positioning device**

99

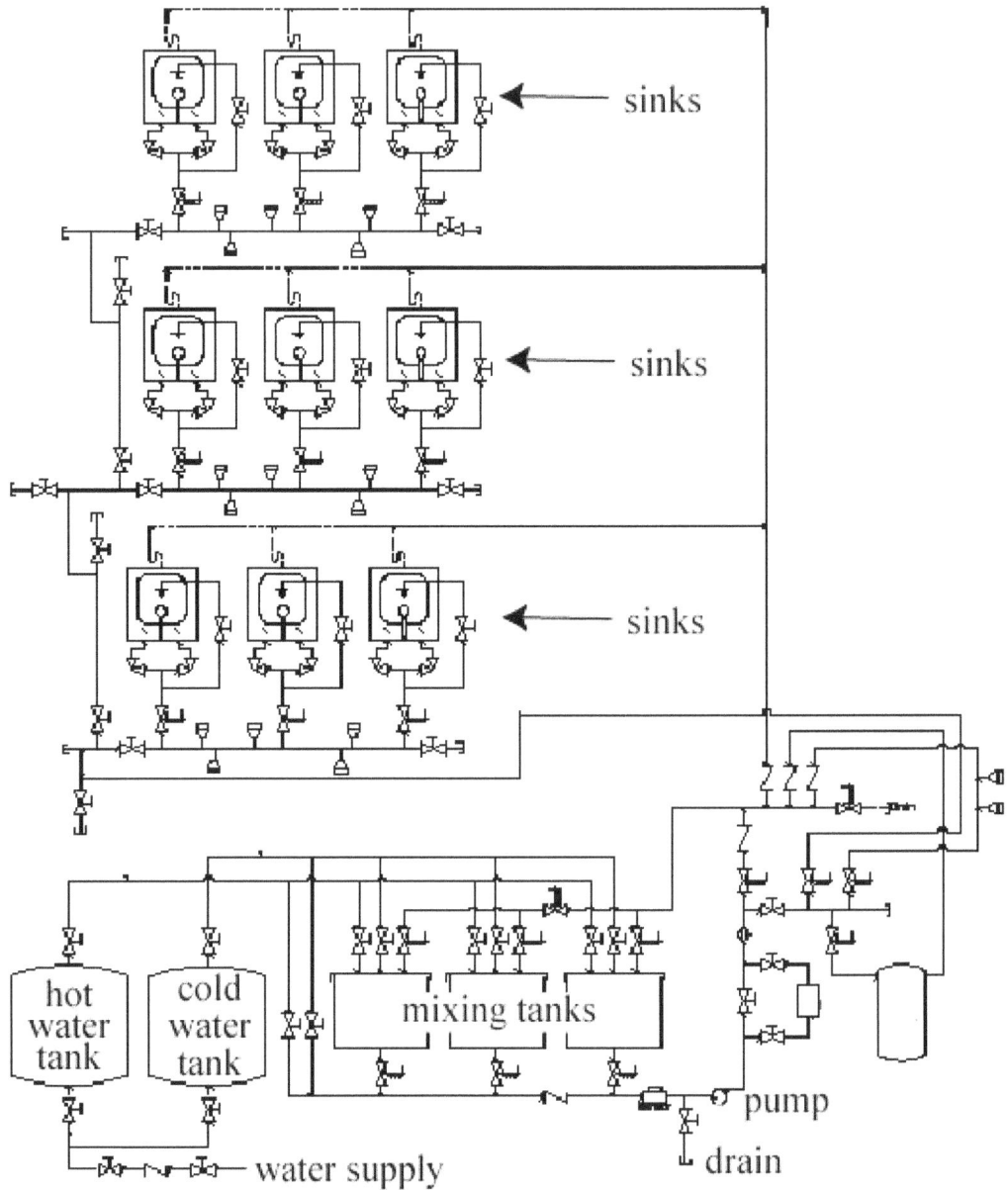

Fig. 3.4.1 Schematic view of the full-scale plumbing system test facility

100

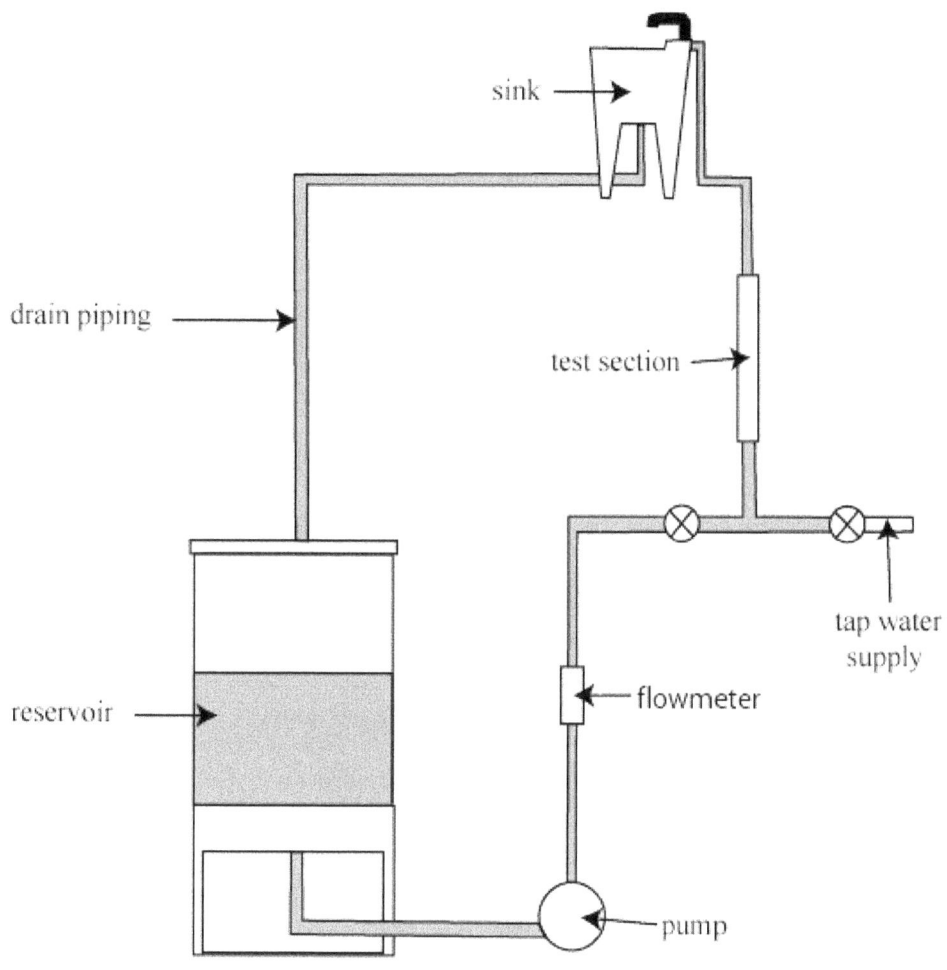

Fig. 3.4.2. Schematic of full-scale test loop

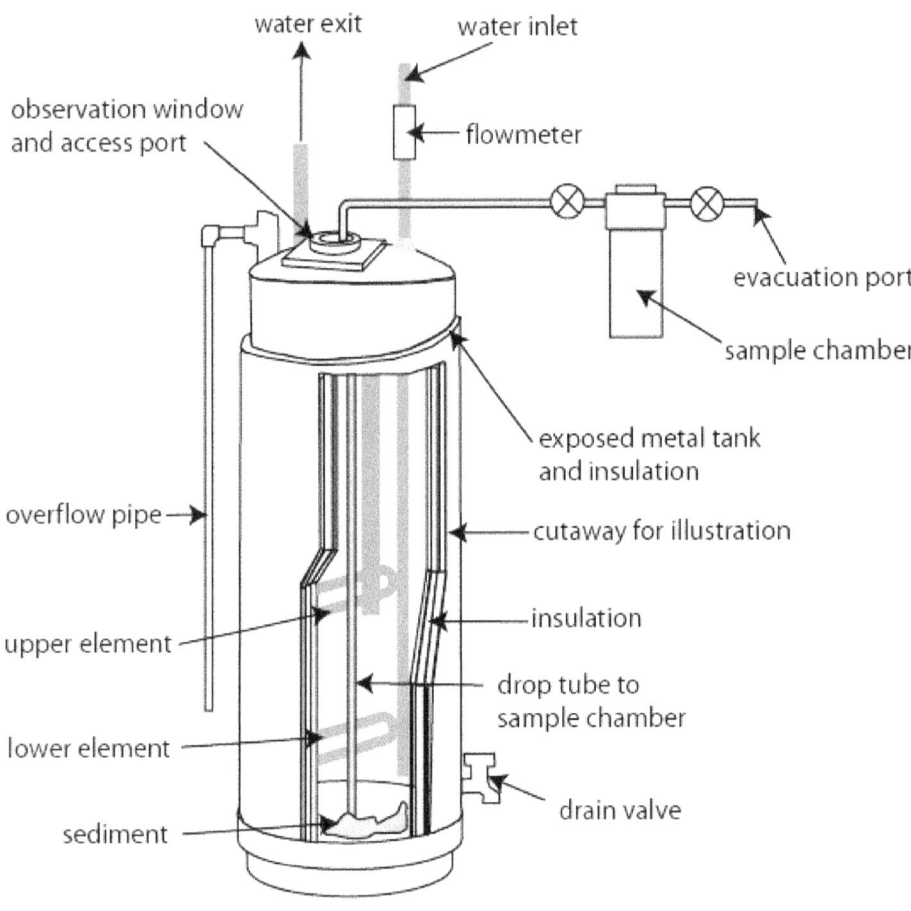

water exit   water inlet

observation window
and access port

flowmeter

evacuation port

sample chamber

exposed metal tank
and insulation

overflow pipe

cutaway for illustration

insulation

upper element

drop tube to
sample chamber

lower element

drain valve

sediment

Fig. 3.4.3. Schematic of hot water heater testing apparatus

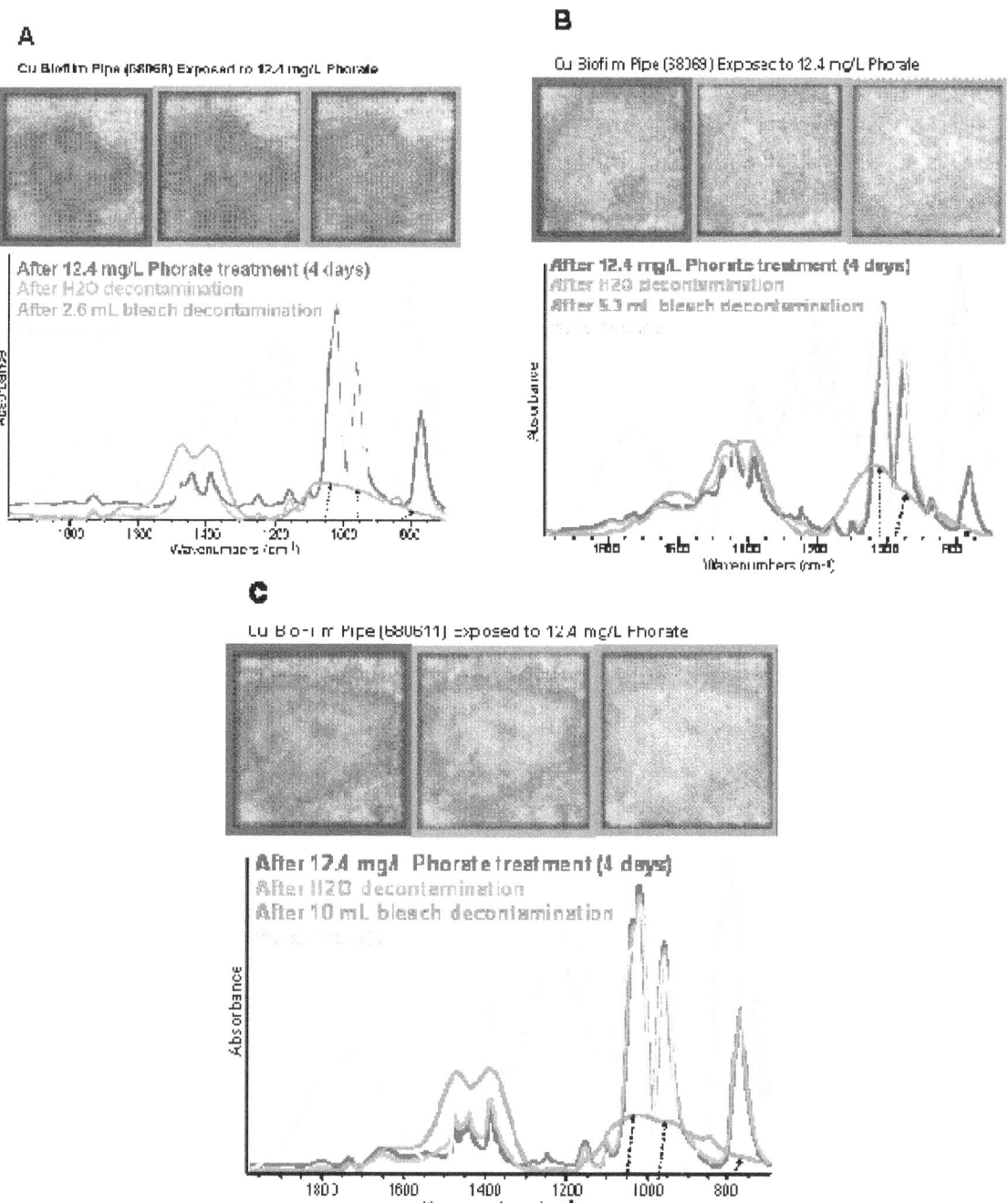

Fig. 4.1.1.1.1 IR micro spectroscopy for Cu biofilm pipes exposed to 12.4 mg/L phorate. Optical images of the Cu pipe are shown above the IR spectra. Figure A represents IR spectra taken on pipe sample 68068 decontaminated in 2.6 mL bleach in water. Figure B represents IR spectra taken on pipe sample 68069 decontaminated in 5.3 mL bleach in water. Figure C represents IR spectra taken on pipe sample 680611 decontaminated in 10 mL bleach in water. Data in blue represent Cu pipe exposed to 12.4 mg/L phorate. Data in pink represent Cu pipe decontaminated in water. Data in red represent Cu pipe decontaminated in the various bleach concentrations. Data in light blue present an IR spectrum of pure phorate. The black arrows show disappearance of the phorate peak.

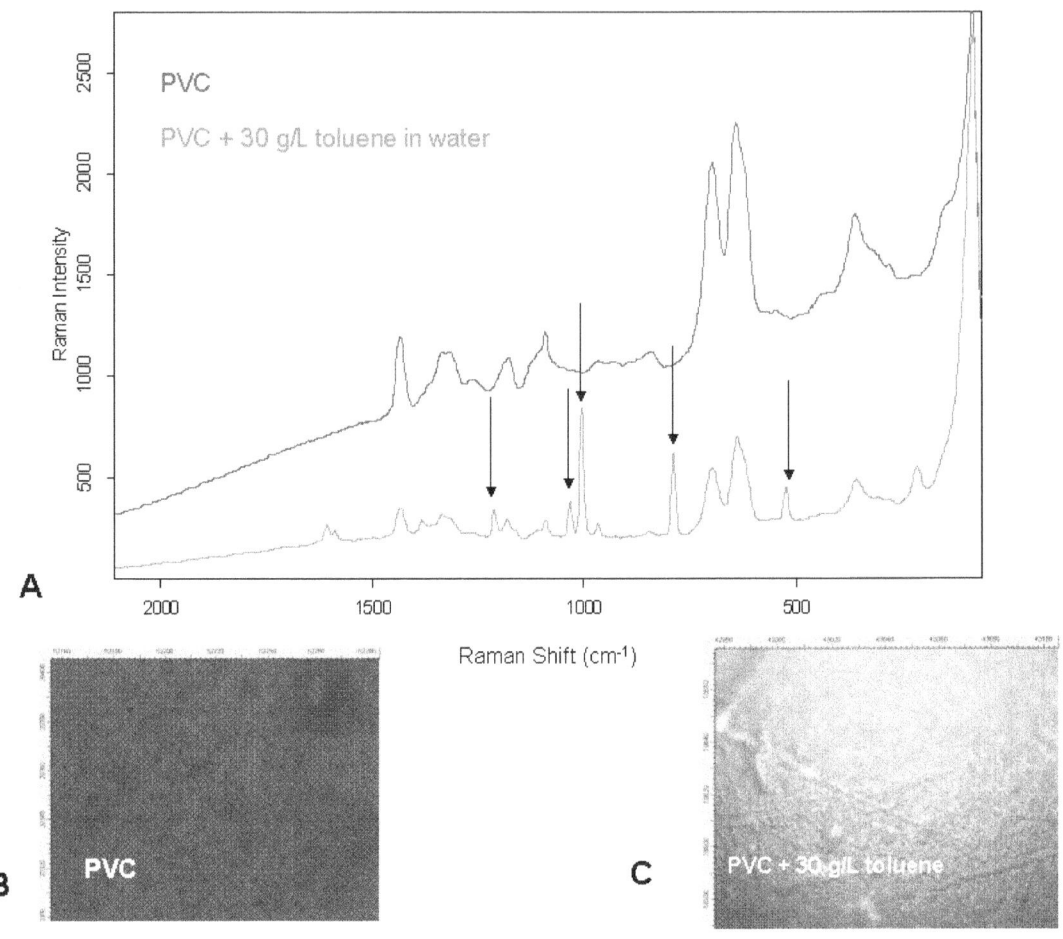

Fig. 4.1.1.2.1 30 g/L toluene in water exposed to PVC pipe. (A) Raman spectra for PVC pipe (blue) and PVC + 30 g/L toluene in water (red). Arrows represent the toluene Raman peaks. (B) and (C) optical images at 50x magnification for PVC and PVC + 30 g/L toluene in water, respectively.

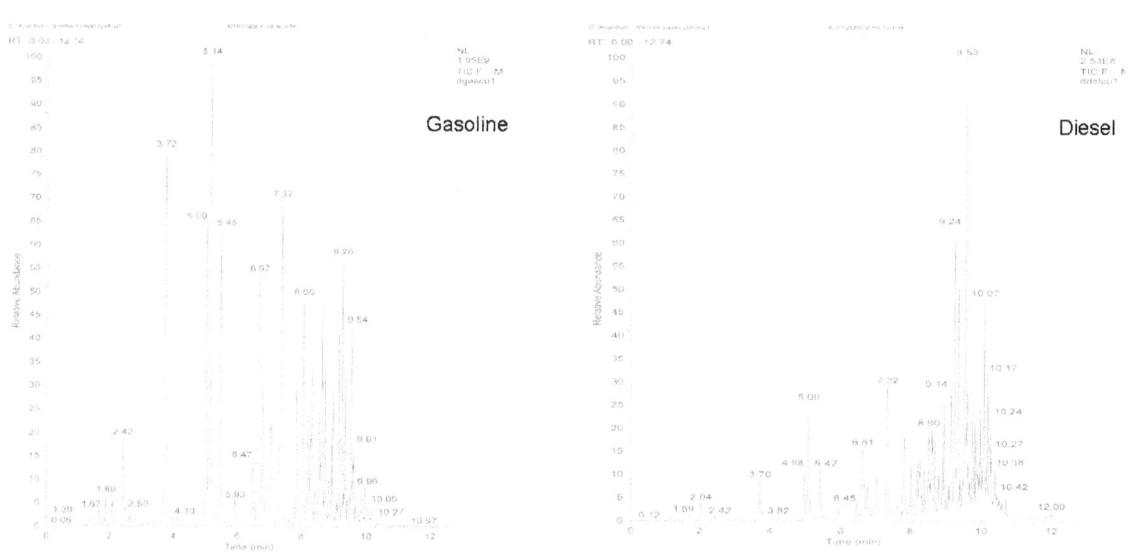

Fig. 4.1.1.3.1 Gas chromatograms for gasoline and diesel

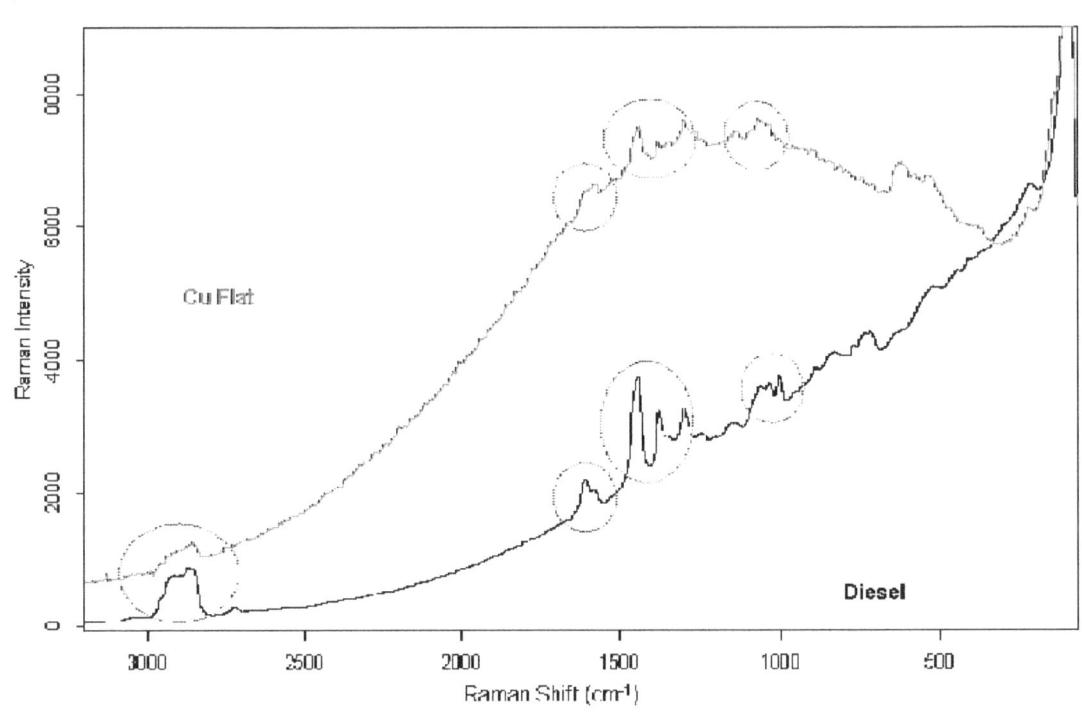

Fig. 4.1.1.3.2 Raman spectra of Cu flat pipe exposed to 2000 mg/L diesel in water. Black spectra represent pure diesel and gray spectra represent diesel on Cu flat pipe. Circled peaks are those peaks from diesel found in both spectra.

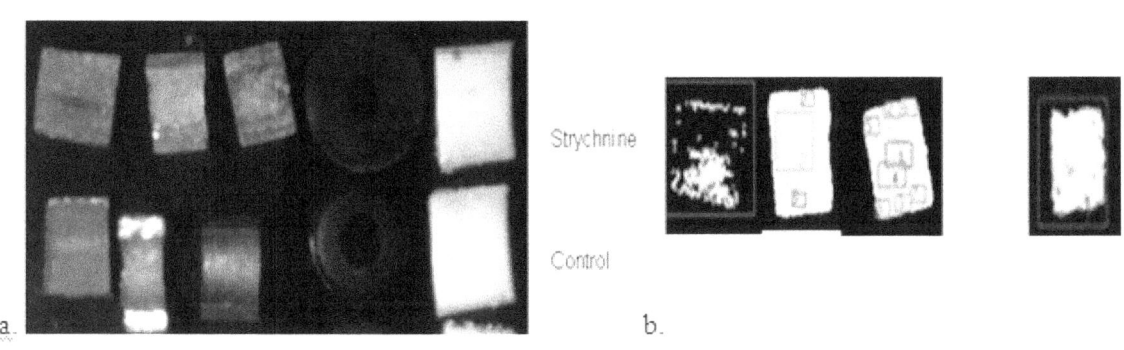

Fig. 4.1.1.4.1 Optical image of control pipes and those exposed to strychnine (a).
Fluorescent images of exposed pipe materials (b). From left to
right, Cu in-service, brass, iron, rubber, and PVC.

Fig. 4.1.1.6.1 SEM image at 2600x magnification of precipitate formed when mercuric chloride in tap water reacts with Cu pipe material. Crystals are copper oxide and droplets are mercury metal.

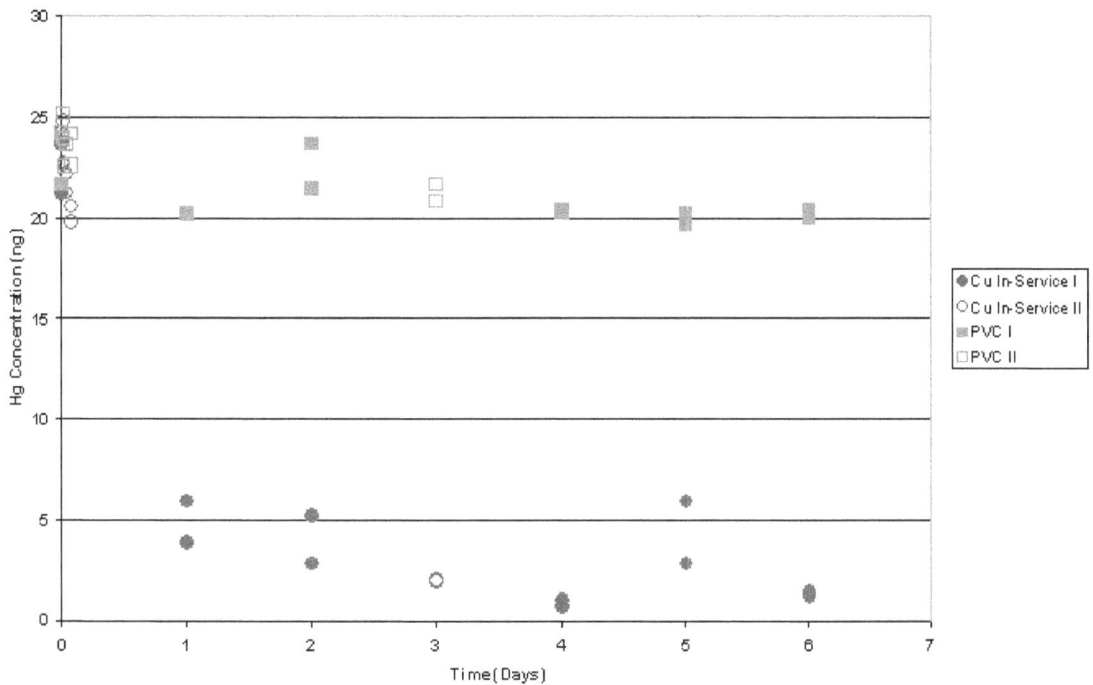

Fig. 4.1.1.6.2 Mercury concentration profile of 500 mg/L HgCl$_2$ solution with PVC (squares) and Cu in-service (circles) pipe materials.

109

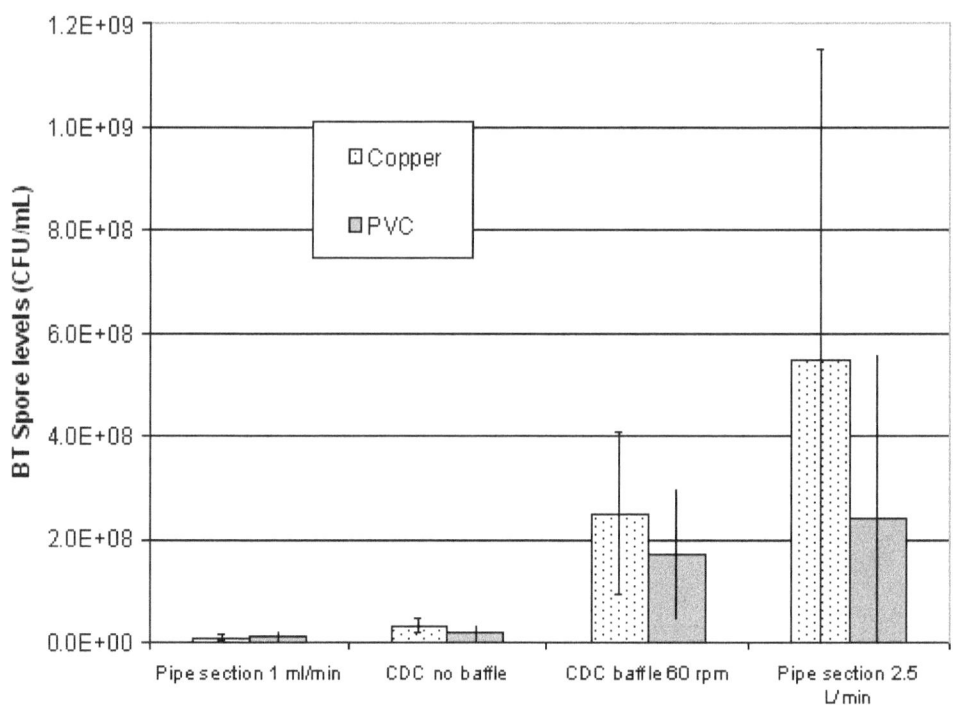

Fig. 4.1.2.2.1 BT Spore levels adhered to biofilm condition pipe surfaces in the reactors with different spore contacting conditions.

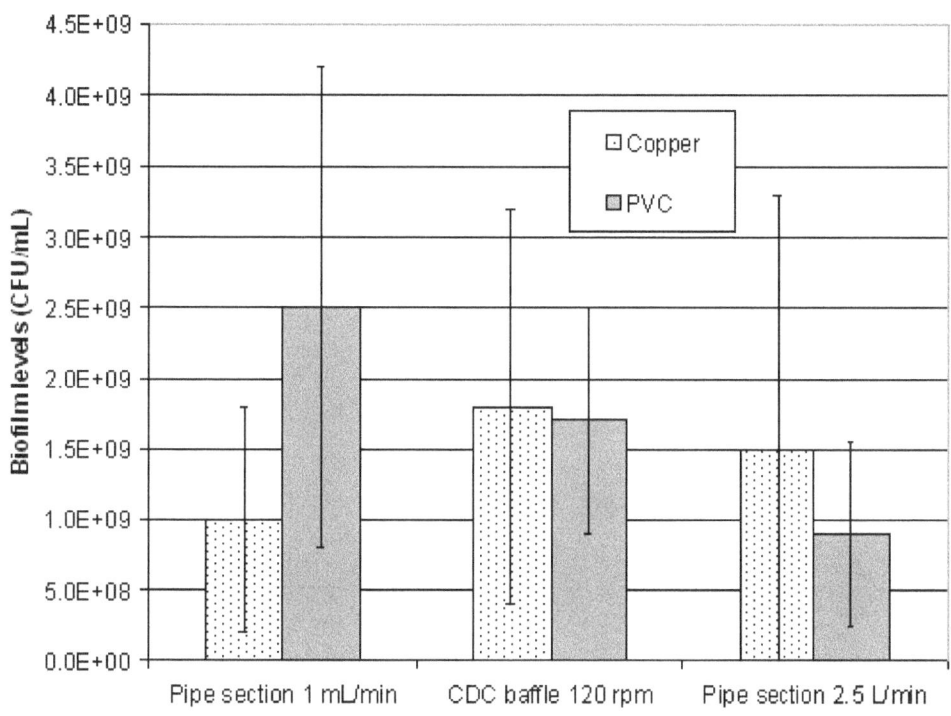

Fig. 4.1.2.1.1 Biofilm organism levels in the reactors with different conditions for growth of the biofilm.

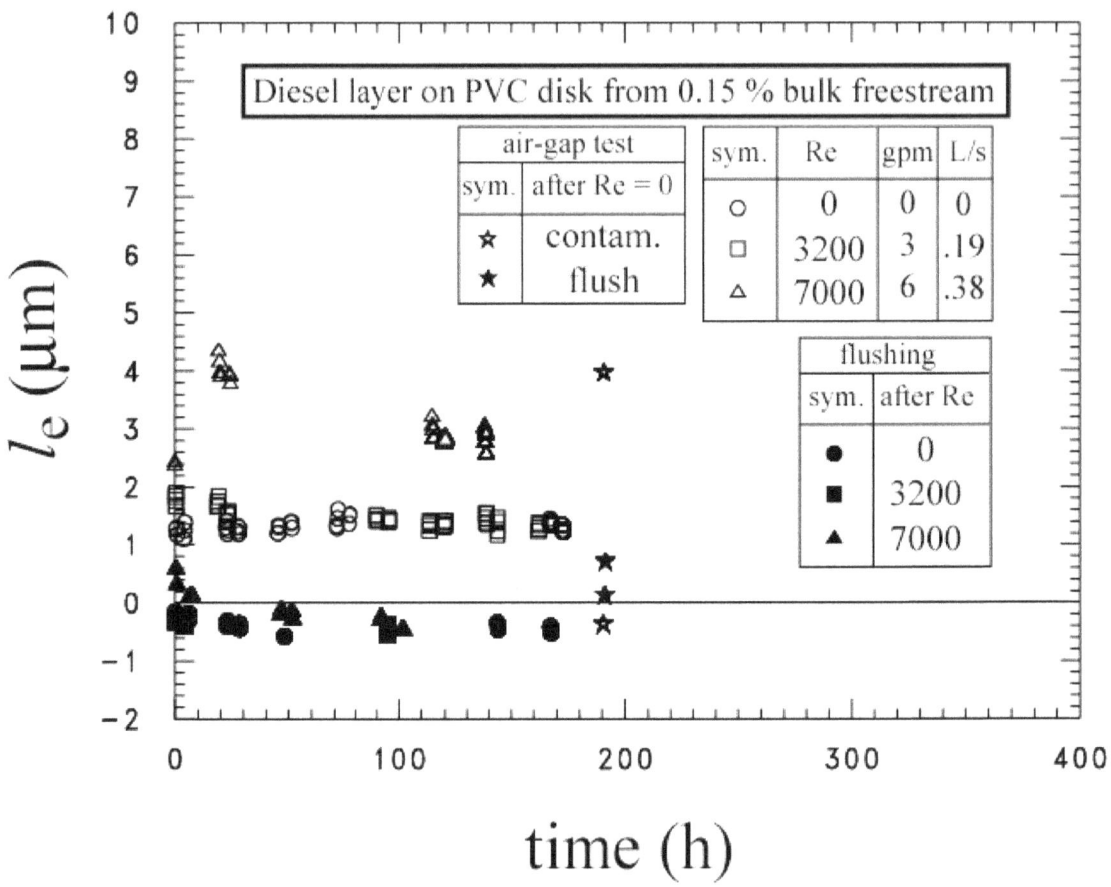

Fig. 4.2.1.1 Effect of exposure time and flow rate on thickness of the diesel excess layer for a 0.15 % bulk freestream mass fraction on a PVC disk

112

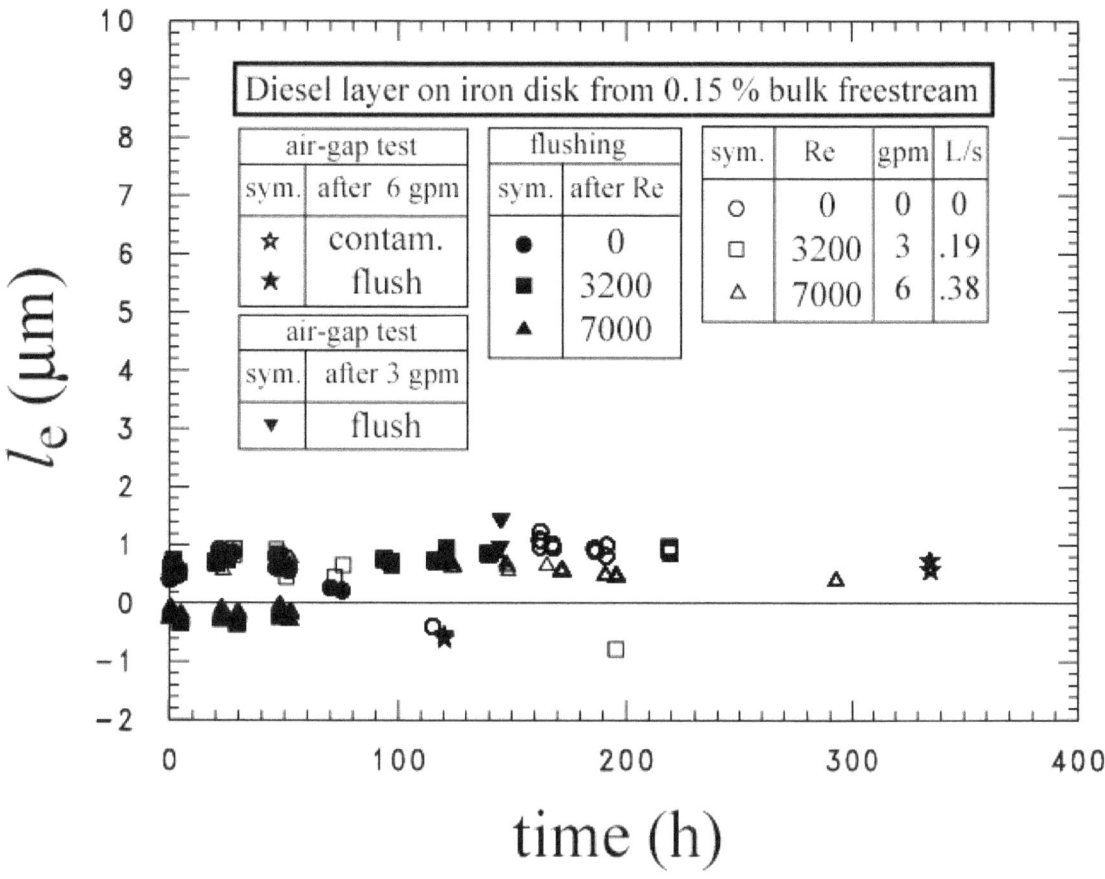

Fig. 4.2.1.2 Effect of exposure time and flow rate on thickness of the diesel excess layer for a 0.15 % bulk free-stream mass fraction on an iron disk

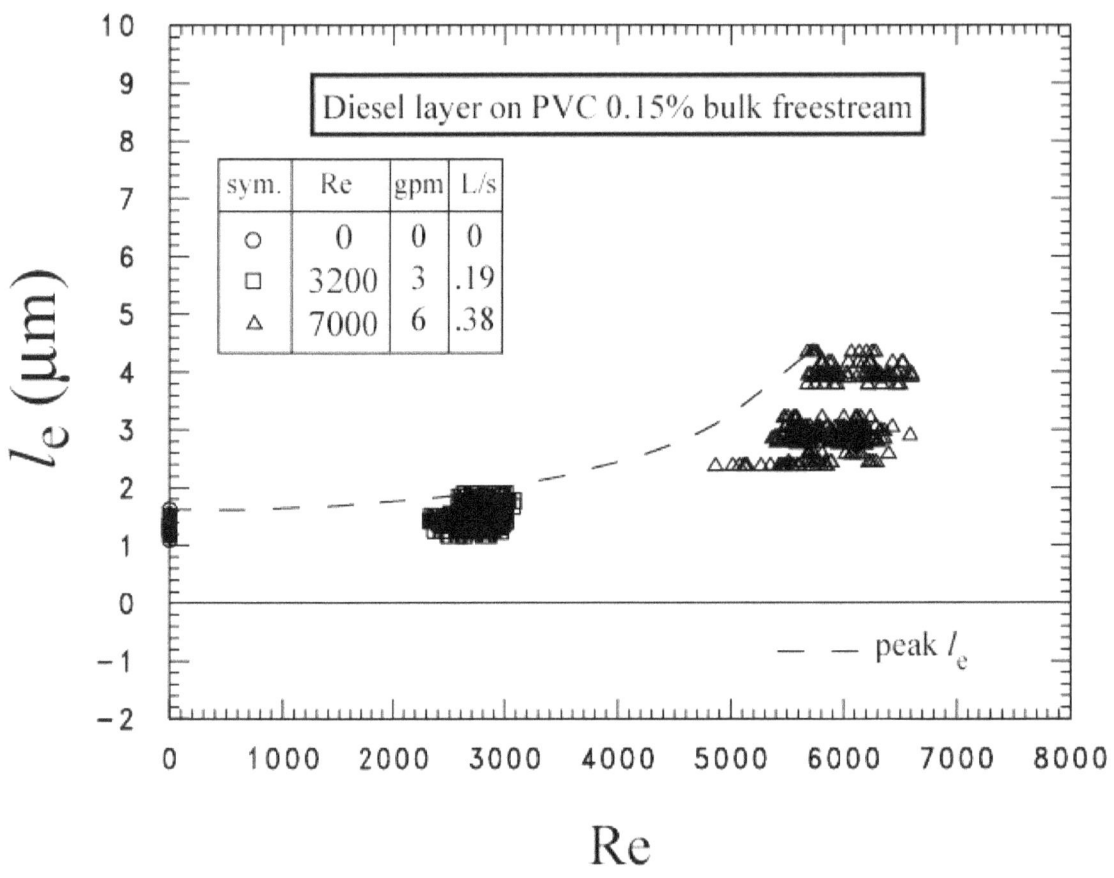

Fig. 4.2.1.3 Diesel excess layer thickness as a function of Re for PVC surface and water/diesel (99.95/0.15)

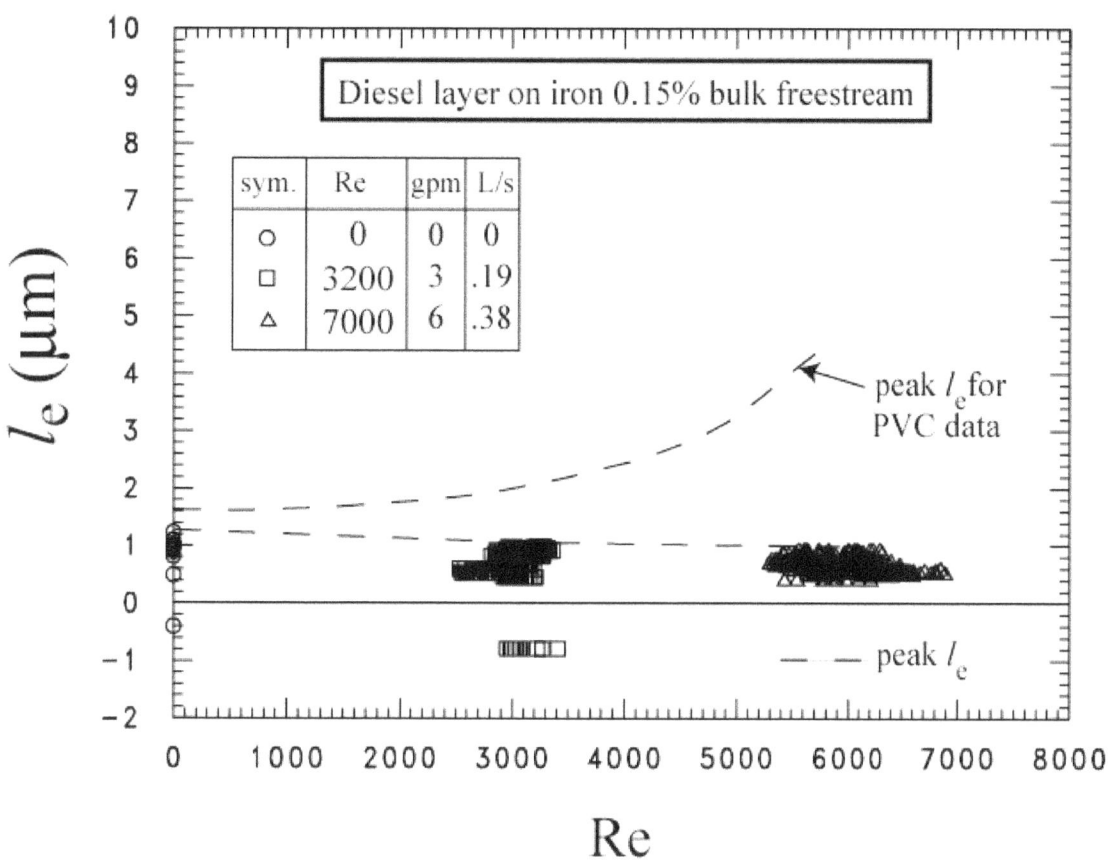

Fig. 4.2.1.4 Diesel excess layer thickness as a function of Re for iron surface and water/diesel (99.95/0.15)

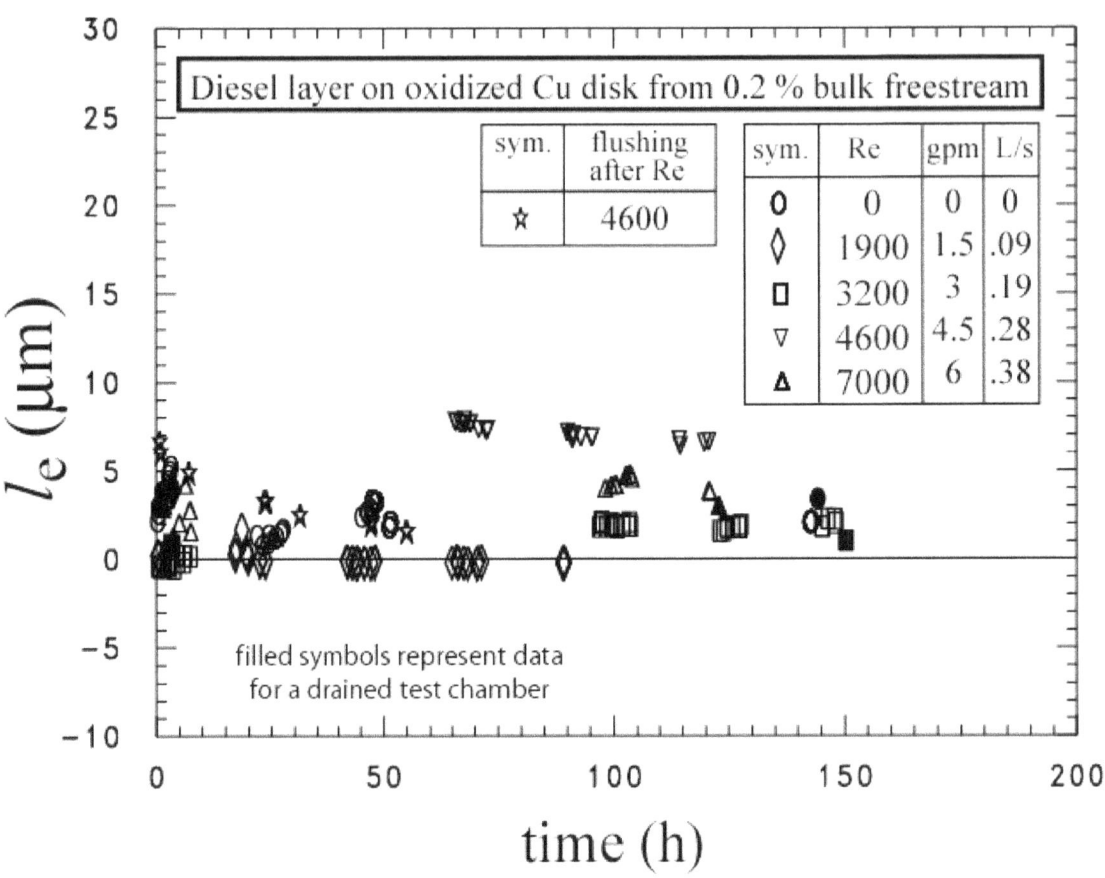

Fig. 4.2.1.5 Effect of exposure time and flow rate on thickness of the diesel excess layer for a 0.2 % bulk free-stream mass fraction

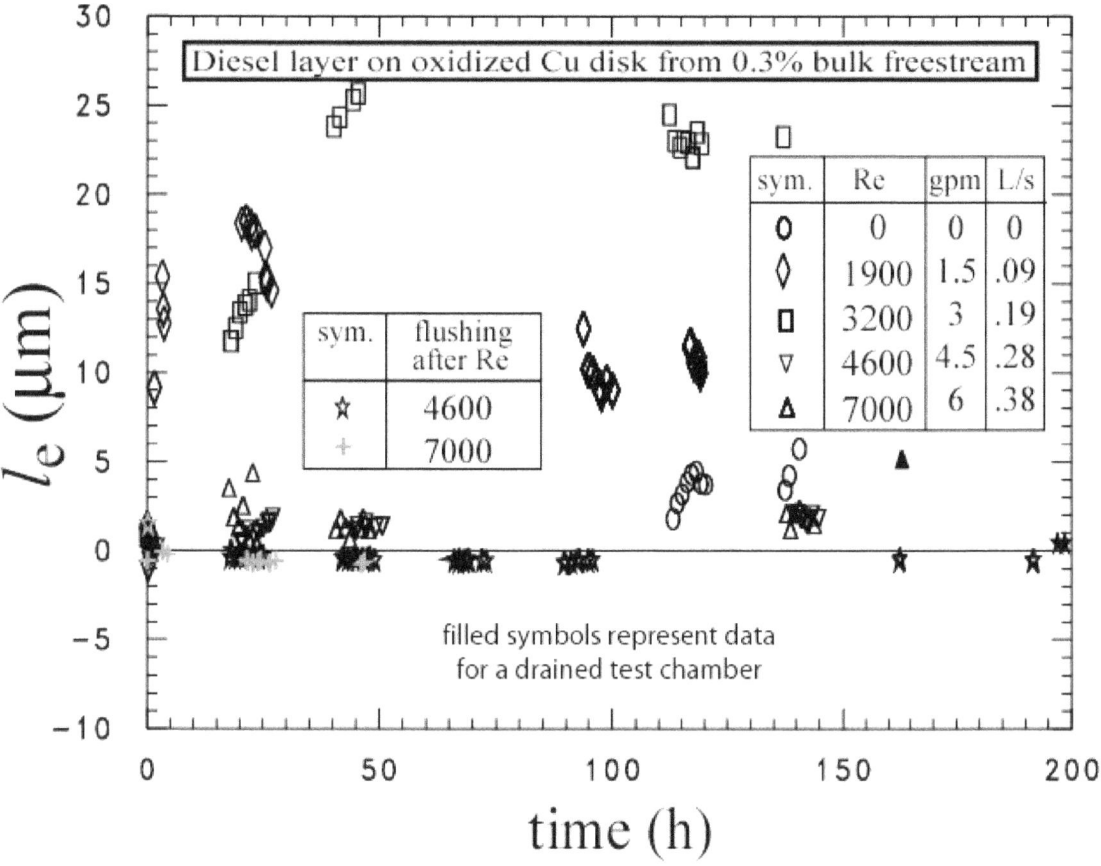

Fig. 4.2.1.6 Effect of exposure time and flow rate on thickness of the diesel excess layer for a 0.3 % bulk free-stream mass fraction

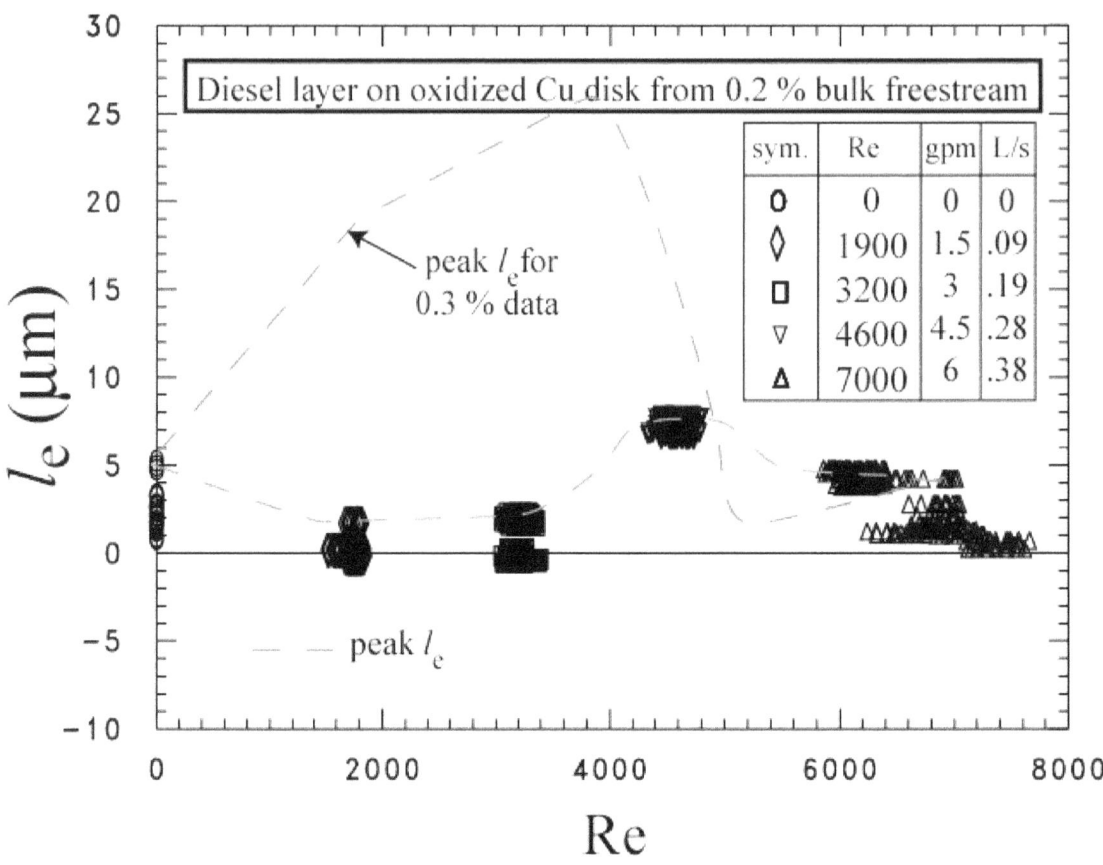

Fig. 4.2.1.7 Diesel excess layer thickness as a function of Re for water/diesel (99.8/0.2)

118

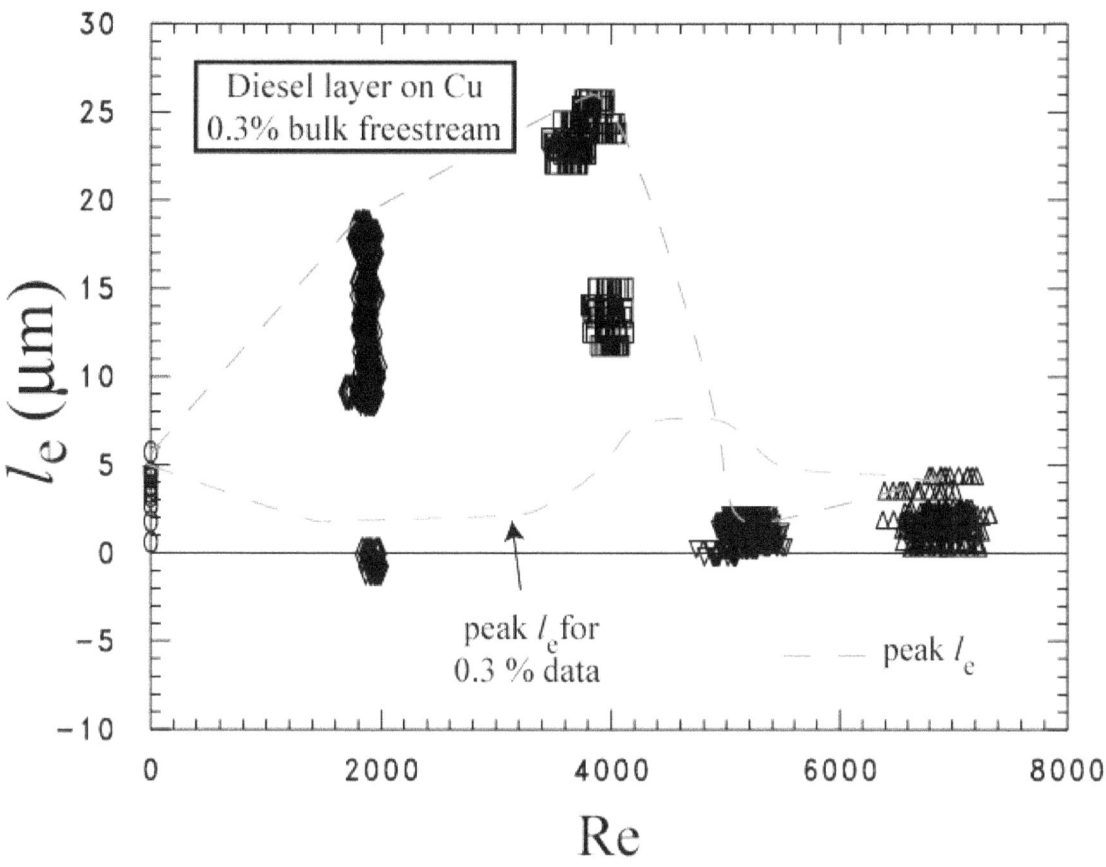

Fig. 4.2.1.8 Diesel excess layer thickness as a function of Re for water/diesel (99.7/0.3)

119

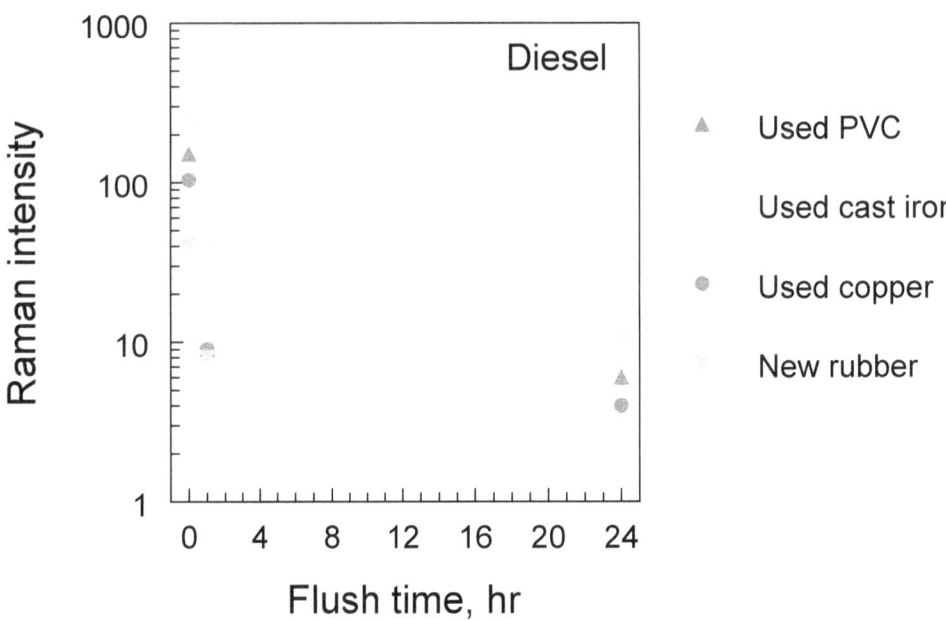

**Fig. 4.3.1. Coupons of copper, PVC, iron, and rubber were soaked in 100 %
diesel fuel for approximately 140 h, except for the rubber, which was
soaked for approximately 283 h. The data points furthest to the left
represent the Raman intensity at a selected wave number for the
specimens immediately after soaking. The data points just to the right
of those points correspond to the specimens having been flushed with
cold water for 1 h. The data points to the far right were taken after
approximately 24 h. of flushing.**

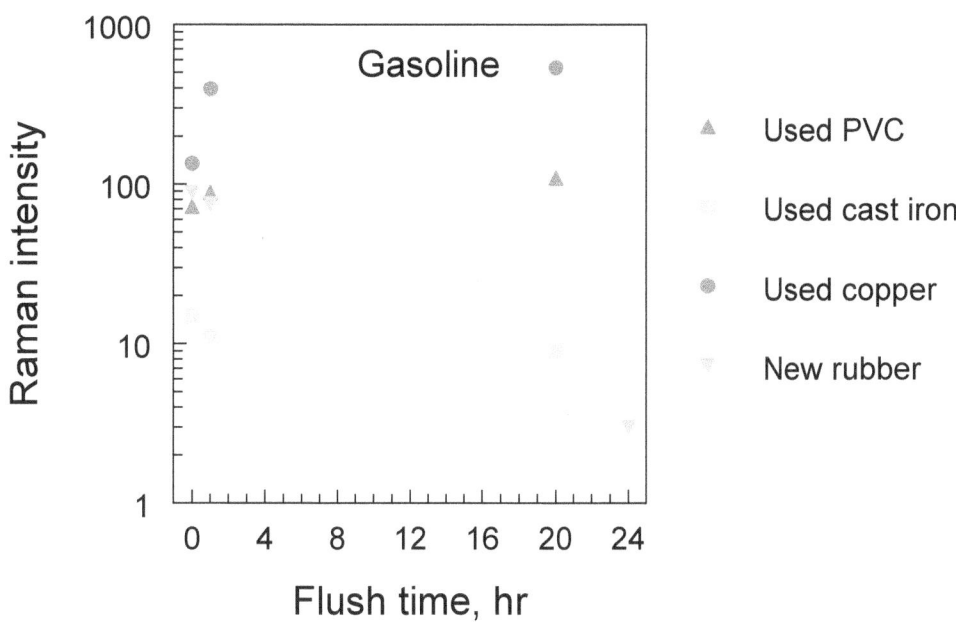

Fig. 4.3.2. Coupons of the materials shown to the right of the graph were soaked
in 100 % gasoline for approximately 24 h, except for the rubber,
which was soaked for approximately 480 h. The data points furthest
to the left represent the Raman intensity at a selected wave number
for the specimens immediately after soaking. The data points just to
the right of those points correspond to the specimens having been
flushed with cold water for 1 h. The data points to the far right were
taken after approximately 20 h. of flushing, except for the rubber
specimen, which was flushed for approximately 24 h.

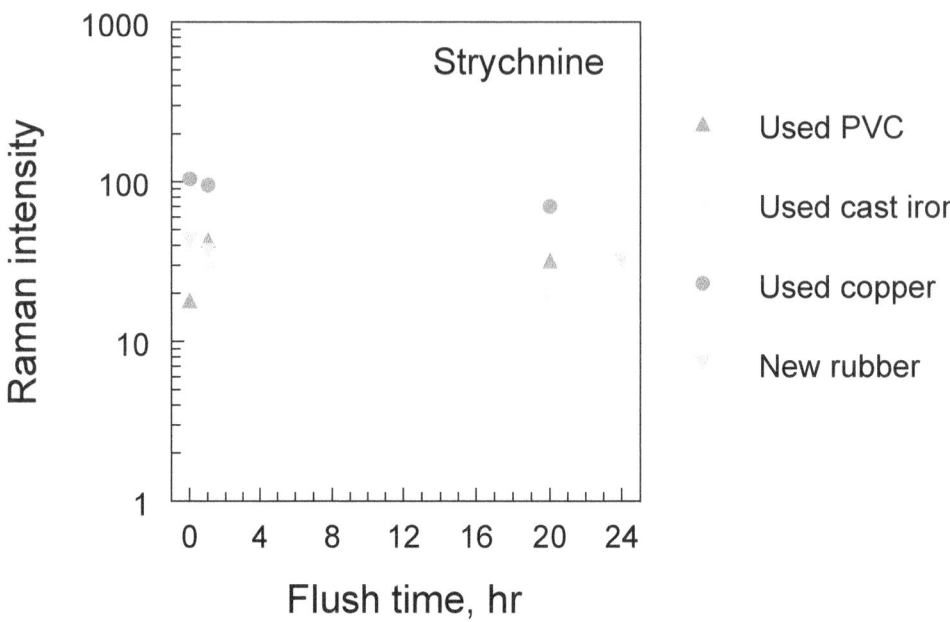

Fig. 4.3.3. Coupons of the materials shown to the right of the graph were soaked in a 0.5 % mass fraction solution (5000 ppm) of strychnine in water for approximately 140 h., except for the rubber, which was soaked for approximately 480 h. The data points furthest to the left represent the Raman intensity at a selected wave number for the specimens immediately after soaking. The data points just to the right of those points correspond to the specimens having been flushed with cold water for approximately 1 h. The data points to the far right were taken after approximately 24 h. of flushing.

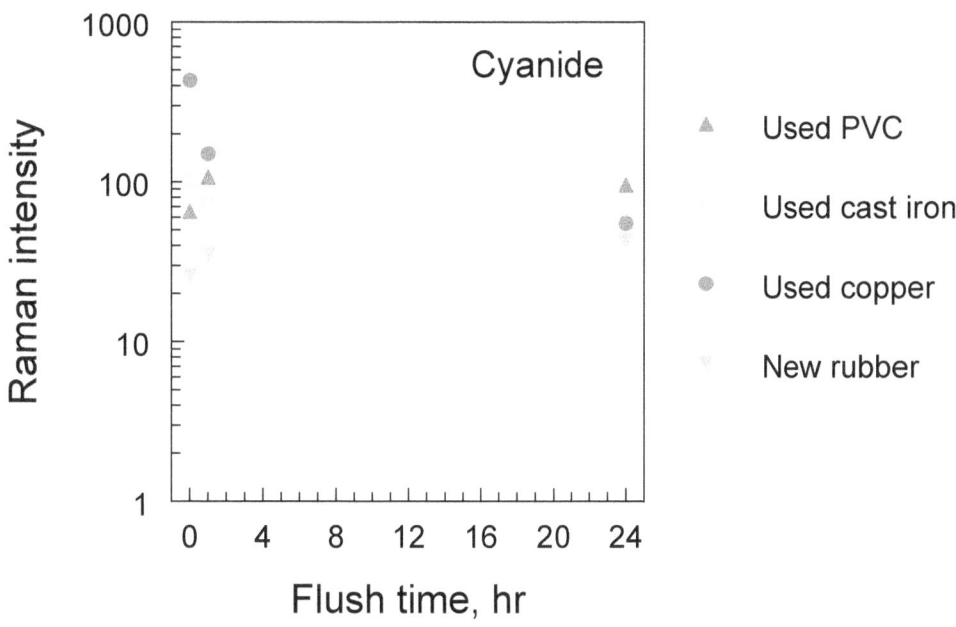

Fig. 4.3.4. Coupons of the materials shown to the right of the graph were soaked in a 1% mass fraction solution (10000 ppm) of sodium cyanide for approximately 260 hr., except for the rubber, which was soaked for approximately 480 hr. The data points furthest to the left represent the Raman intensity at a selected wave number for the specimens immediately after soaking. The data points just to the right of those points correspond to the specimens having been flushed with cold water for approximately 1 hr. The data points to the far right were taken after approximately 24 hr. of flushing.

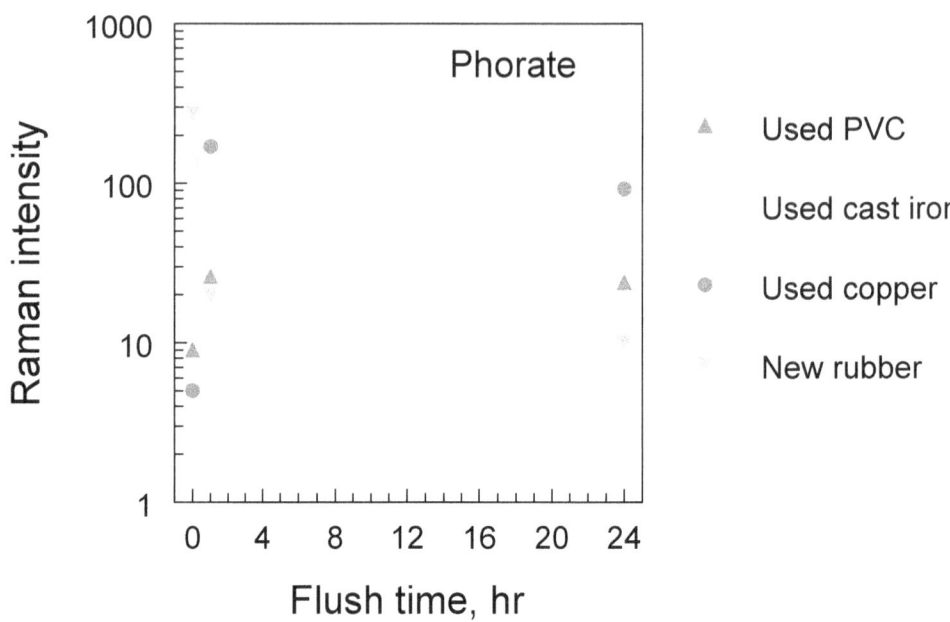

Fig. 4.3.5. Coupons of the materials shown to the right of the graph were soaked in a solution of 100% phorate (a systemic insecticide, $C_7H_{17}O_2PS_3$) for approximately 144 h, except for the rubber, which was soaked for approximately 283 h. The data points furthest to the left represent the Raman intensity at a selected wave number for the specimens immediately after soaking. The data points just to the right of those points correspond to the specimens having been flushed with cold water for approximately 1 h. The data points to the far right were taken after 24 h. of flushing.

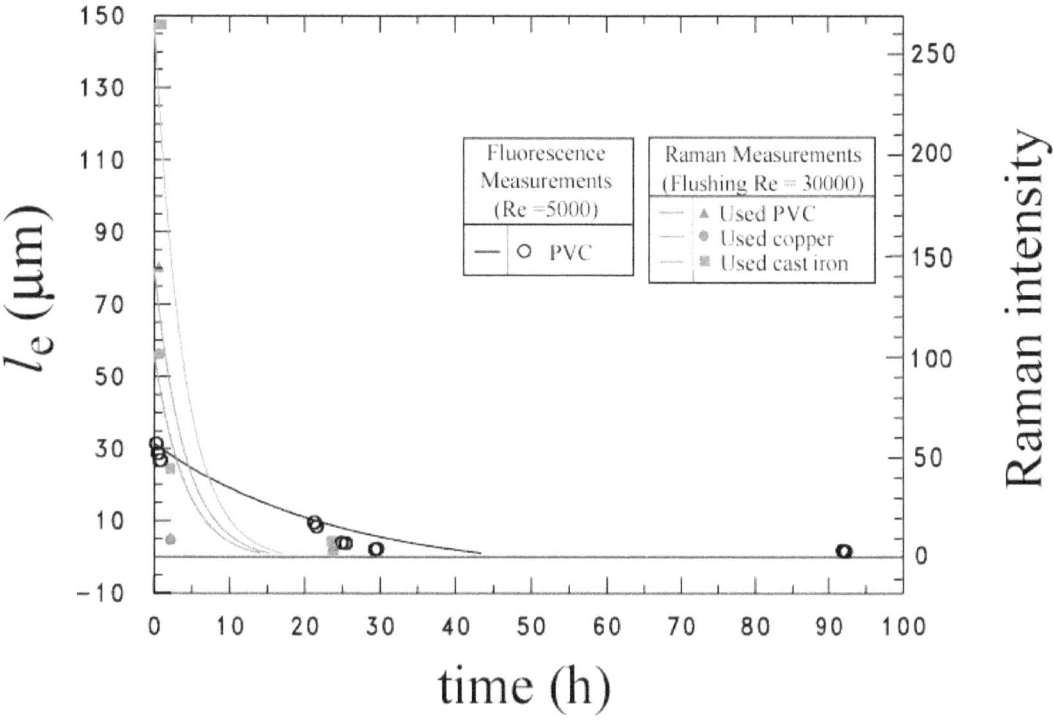

Fig. 4.3.6. **Comparison of flushing test results using the fluorescent measurement technique and the Raman measurement technique for different exposure conditions and Reynolds numbers**

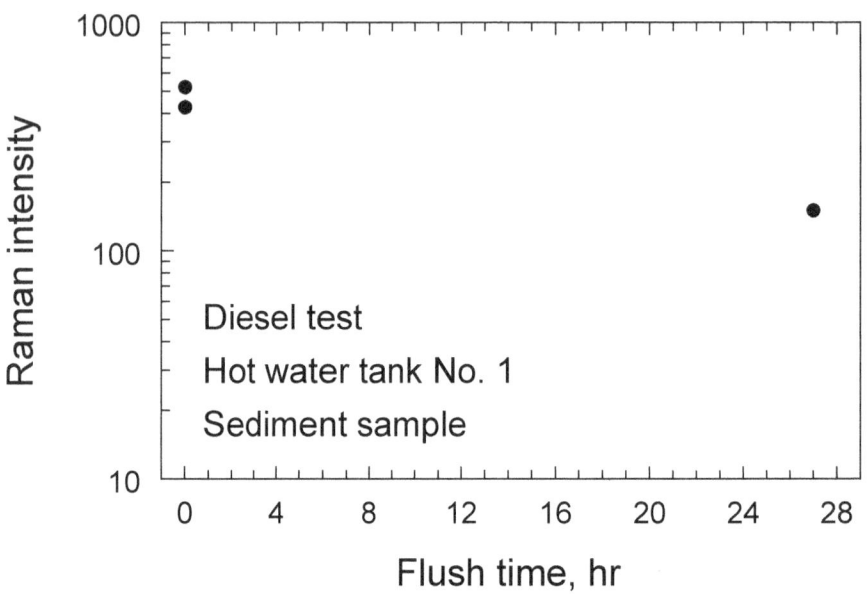

Fig. 4.4.1.1 Measurement results for a hot water tank exposed to diesel fuel

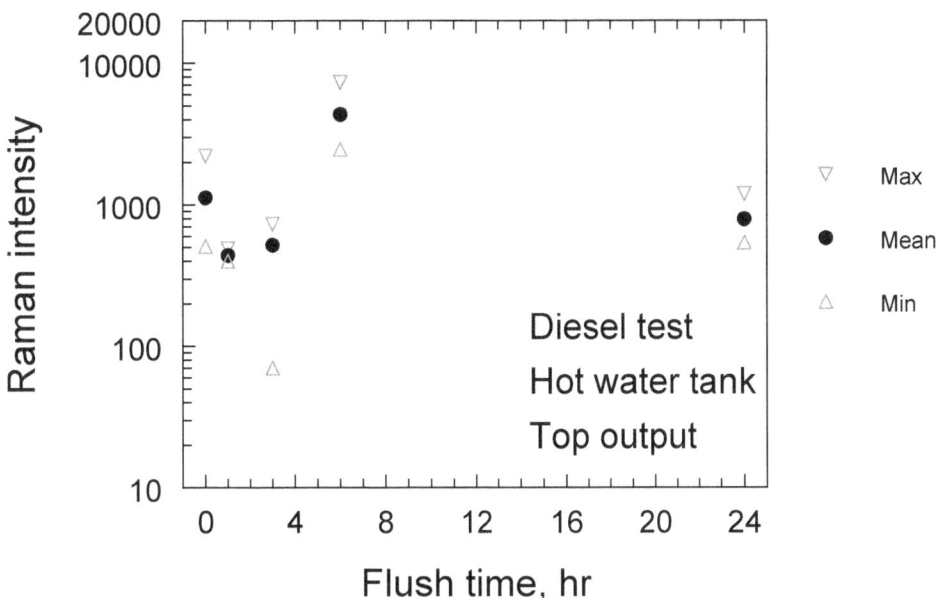

Fig. 4.4.1.2 Heated hot water heater test with diesel fuel, sampling from outlet

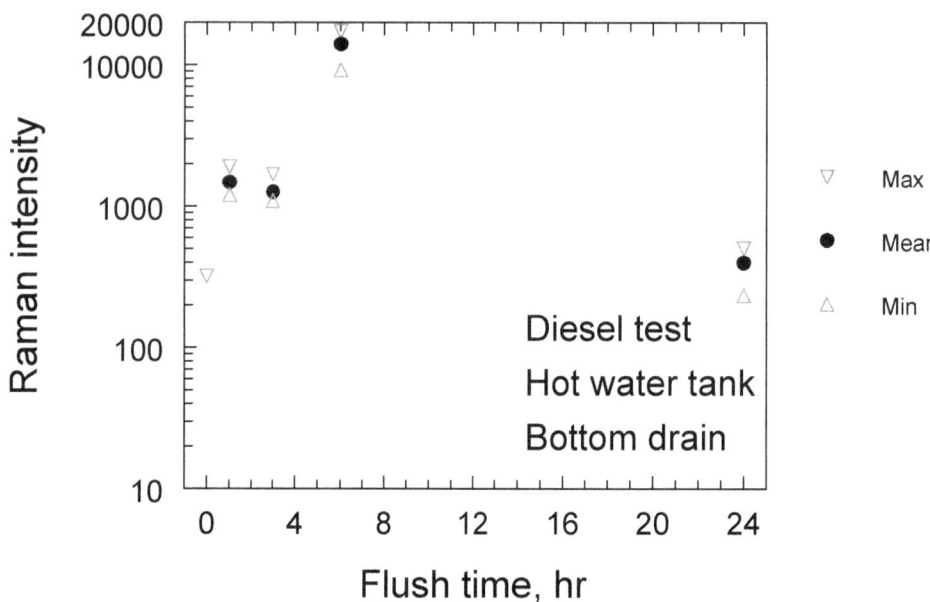

Fig. 4.4.1.3 Heated hot water heater test with diesel fuel, sampling from drain

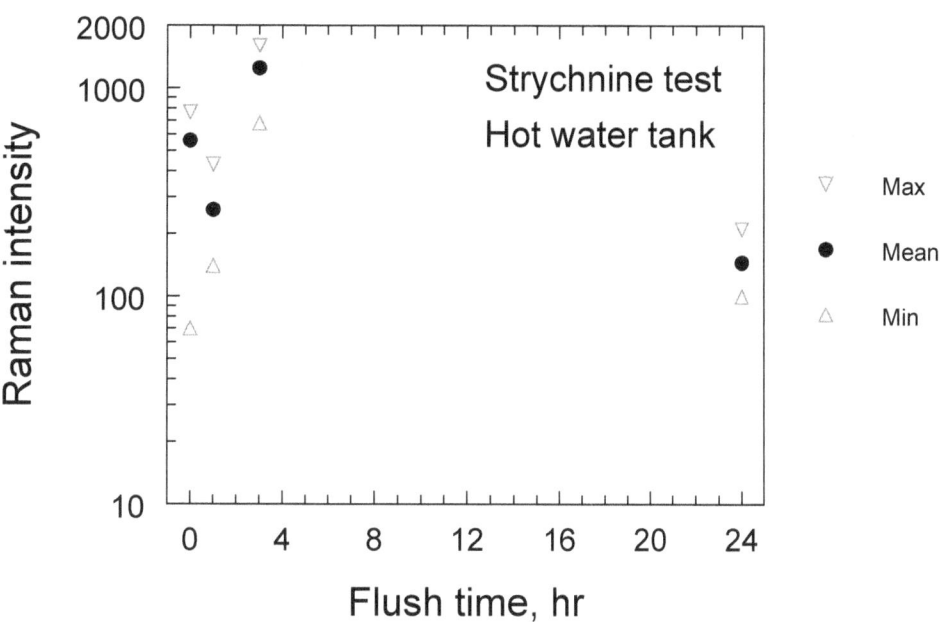

Fig. 4.4.2 Hot water heater test with strychnine

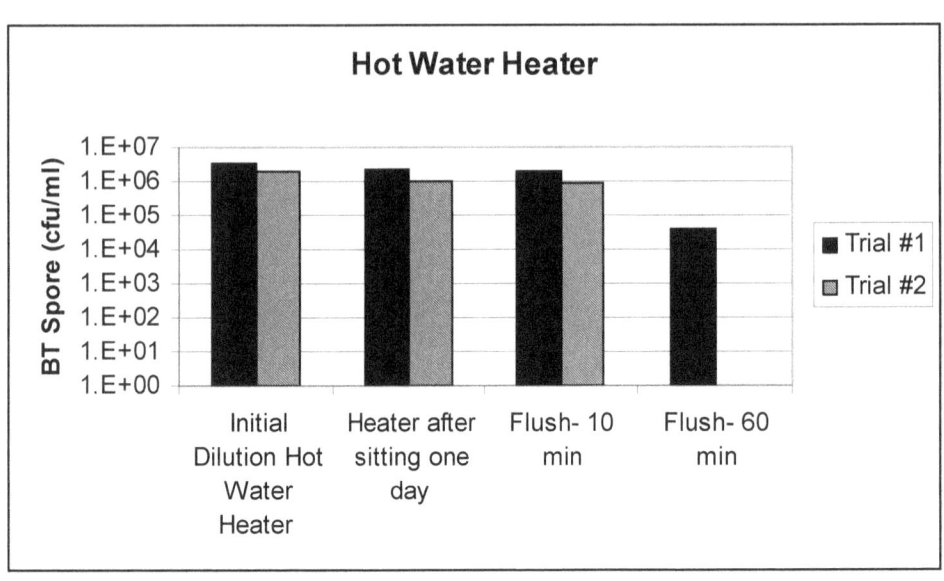

Fig. 4.4.4.1 Hot water heater measurements with BT spores- water samples

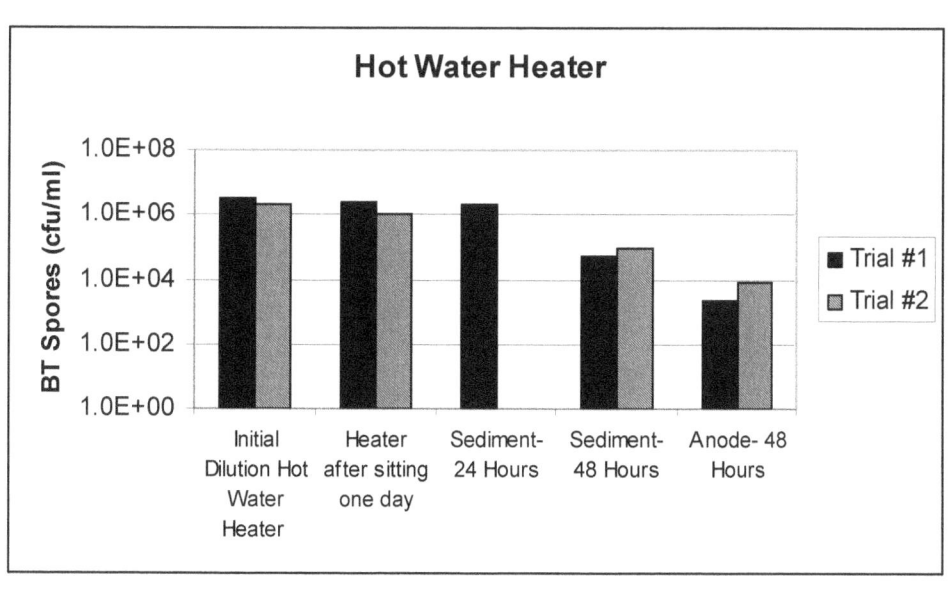

Fig. 4.4.4.2 Hot water heater measurements with BT spores sediment and anode samples

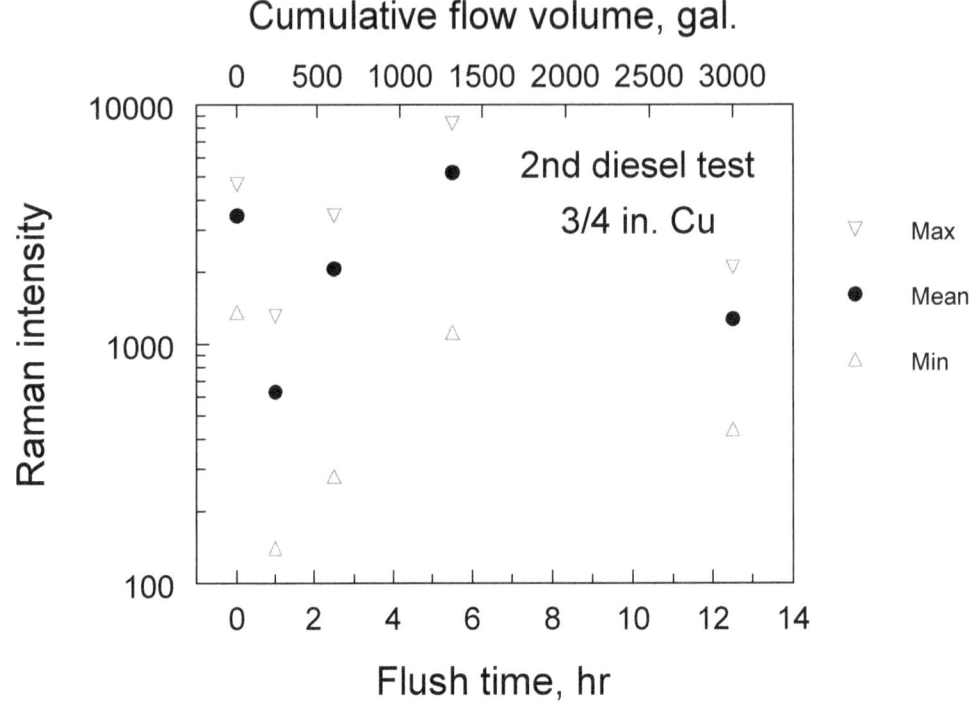

Fig. 4.5.1.1 Pipe loop measurements for copper pipe and diesel fuel

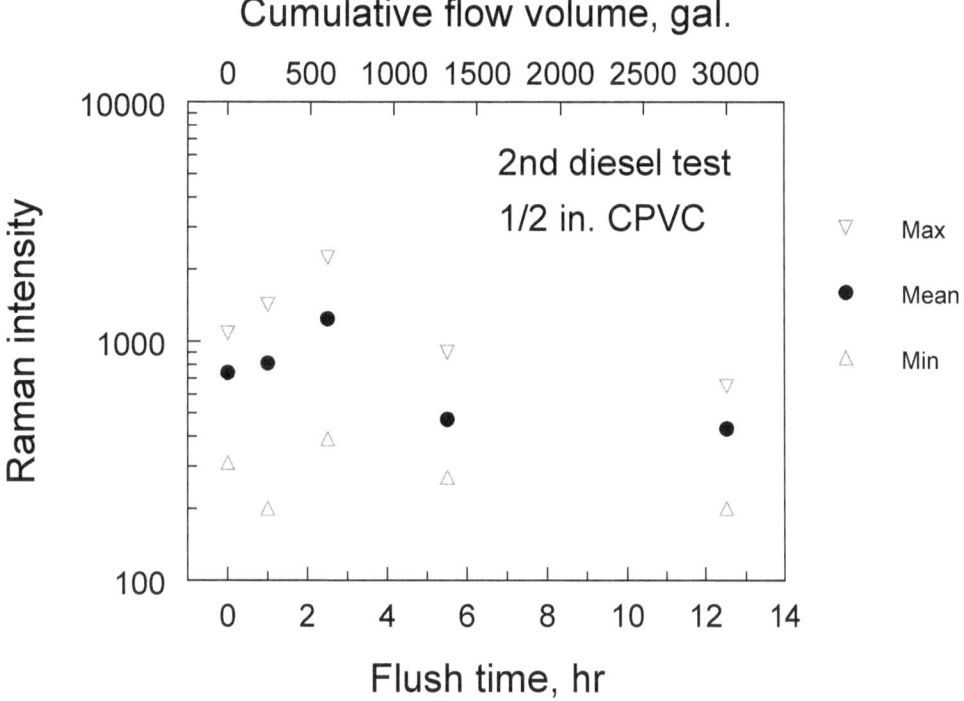

Fig. 4.5.1.2 Pipe loop measurements for CPVC pipe and diesel fuel

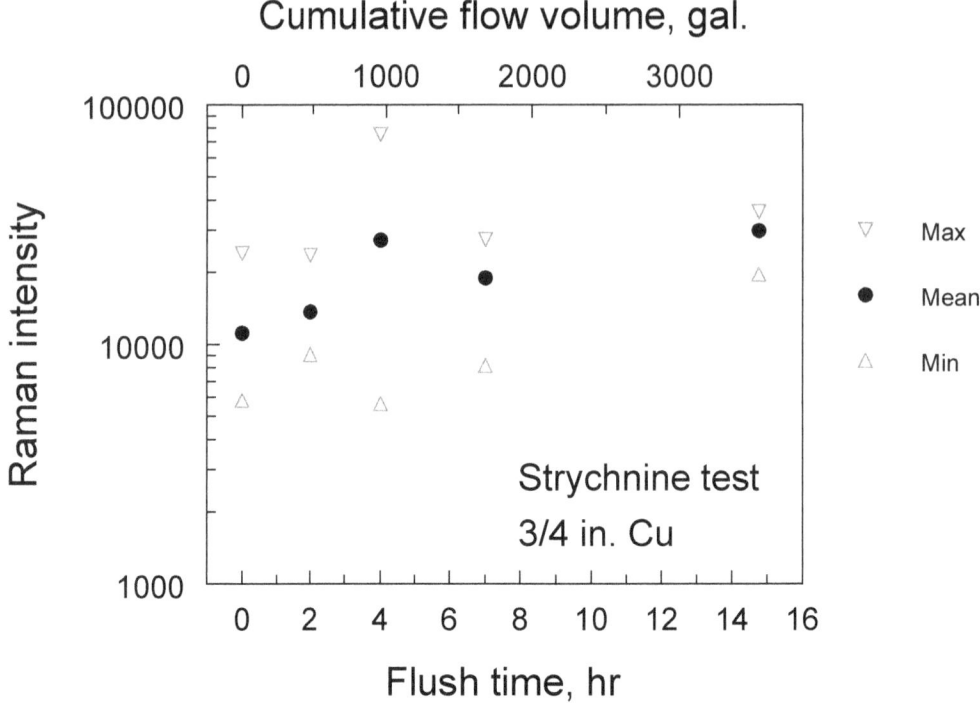

Fig. 4.5.2.1 Pipe loop measurements for ¾ inch copper pipe and strychnine

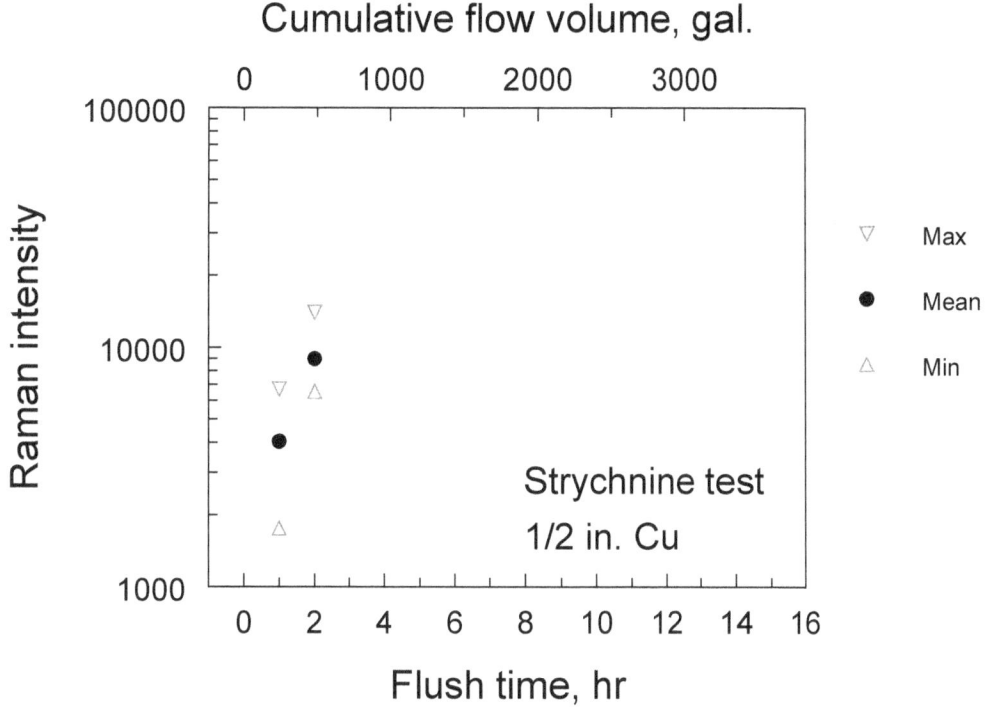

Fig. 4.5.2.2 Pipe loop measurements for ½ inch copper pipe and strychnine

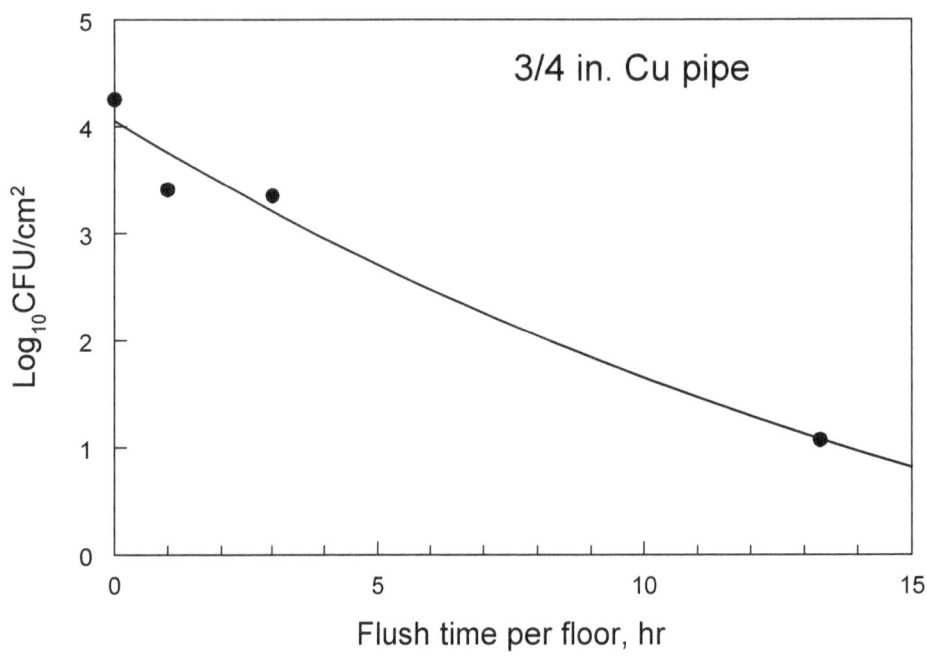

Fig. 4.5.3.1 Pipe loop measurements for ¾ inch copper pipe and BT spores

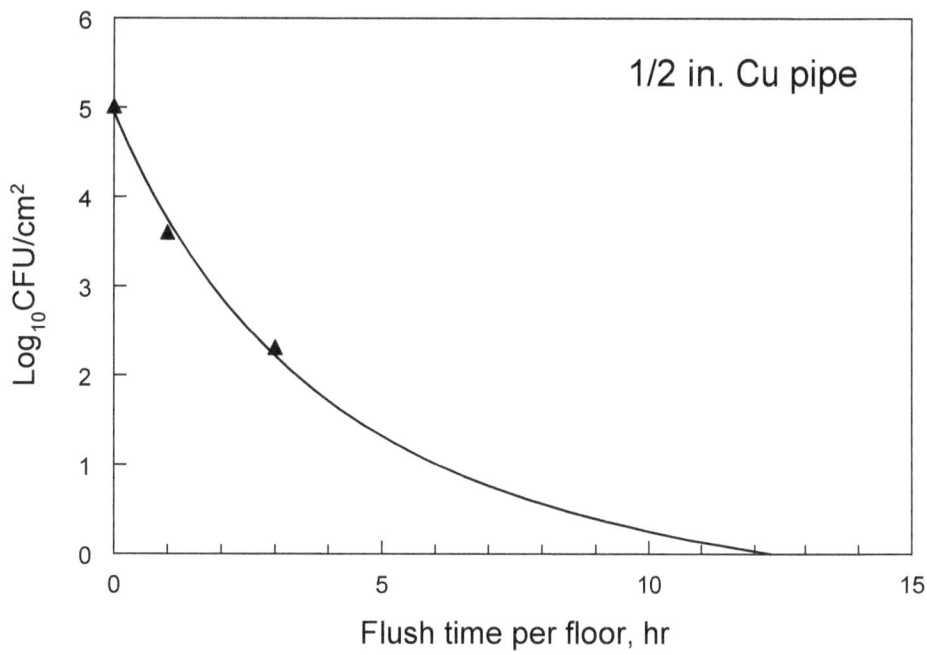

Fig. 4.5.3.2 Pipe loop measurements for ½ inch copper pipe and BT spores

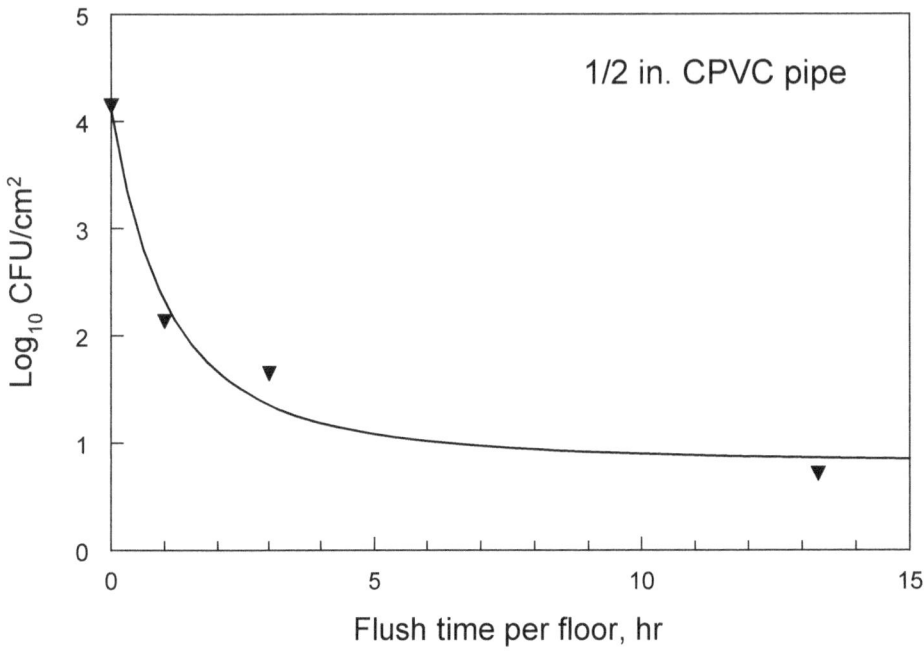

Fig. 4.5.3.3 Pipe loop measurements for ½ inch CPVC pipe and BT spores

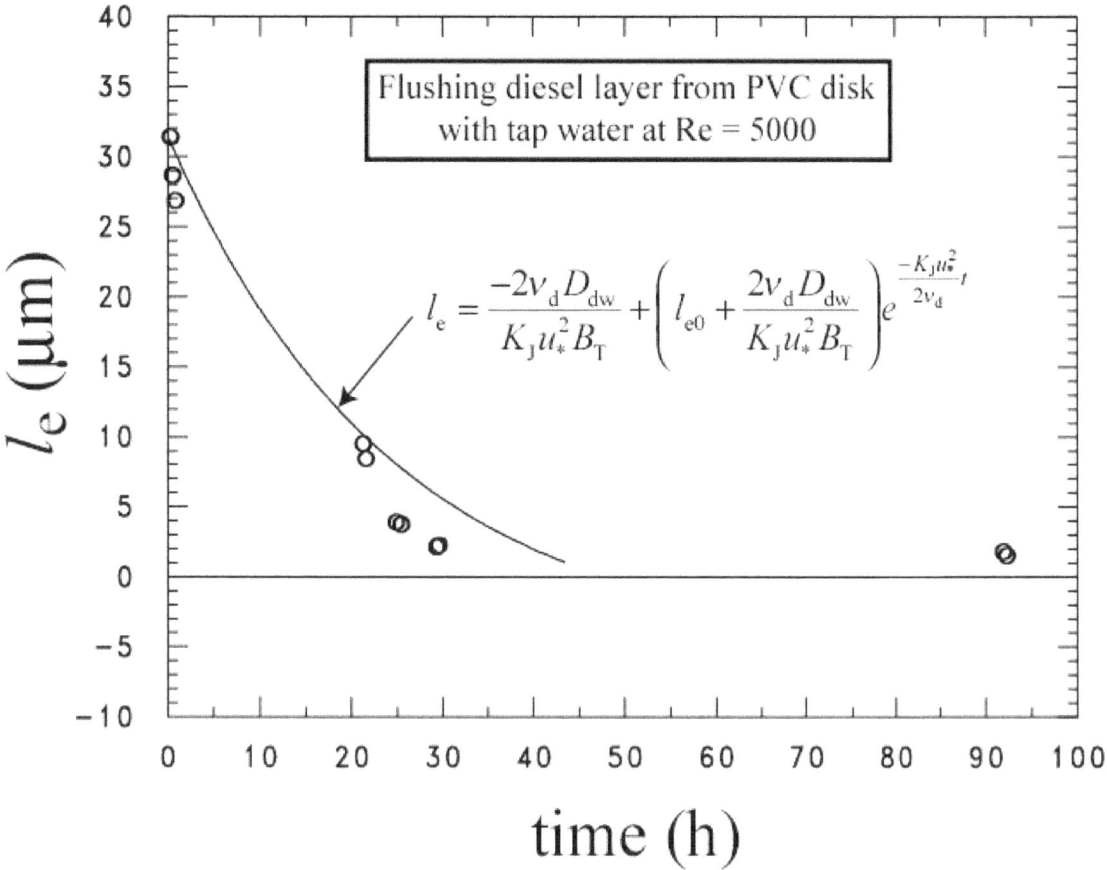

$$l_e = \frac{-2v_d D_{dw}}{K_J u_*^2 B_T} + \left(l_{e0} + \frac{2v_d D_{dw}}{K_J u_*^2 B_T} \right) e^{\frac{-K_J u_*^2}{2v_d}t}$$

Fig. 5.1.1.1 Flushing measurements used to fit coefficients of model

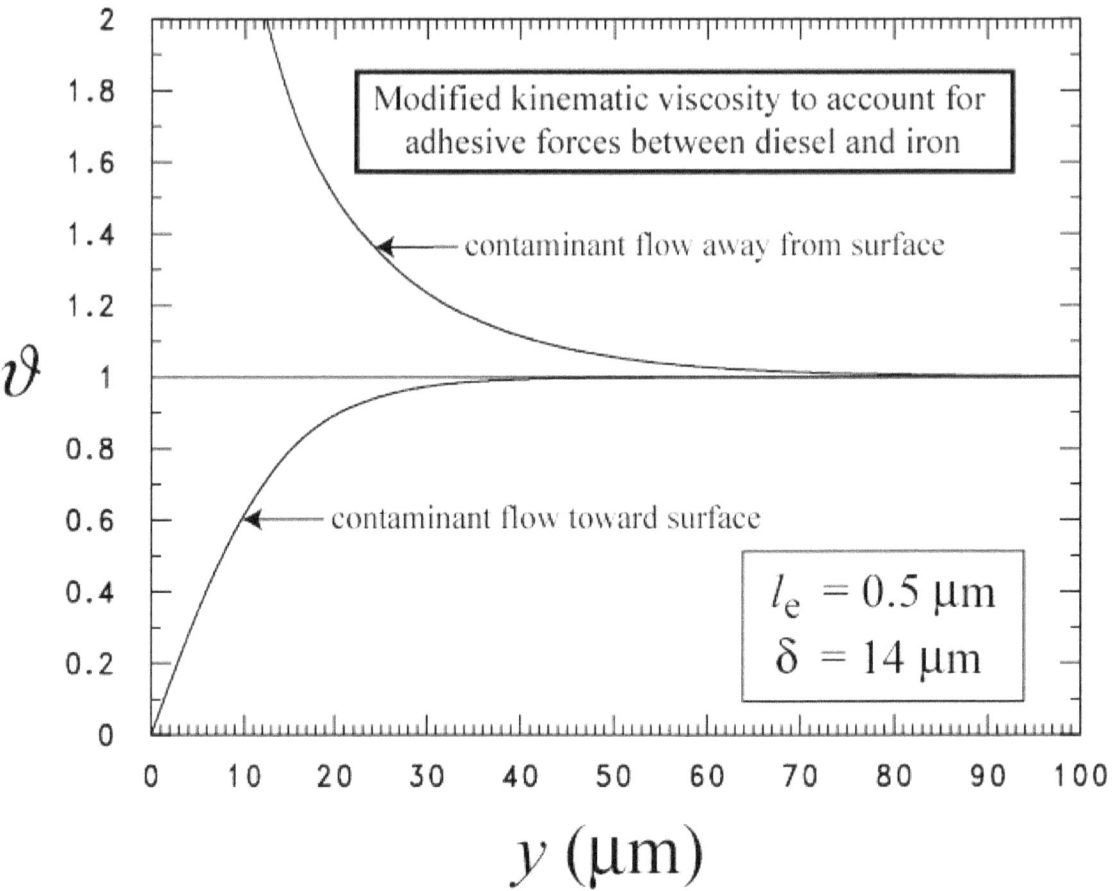

Fig. 5.1.2.1 Modified kinematic viscosity to account for adhesive forces between iron and diesel derived for two flow conditions (Re = 3200 and Re = 7000)

140

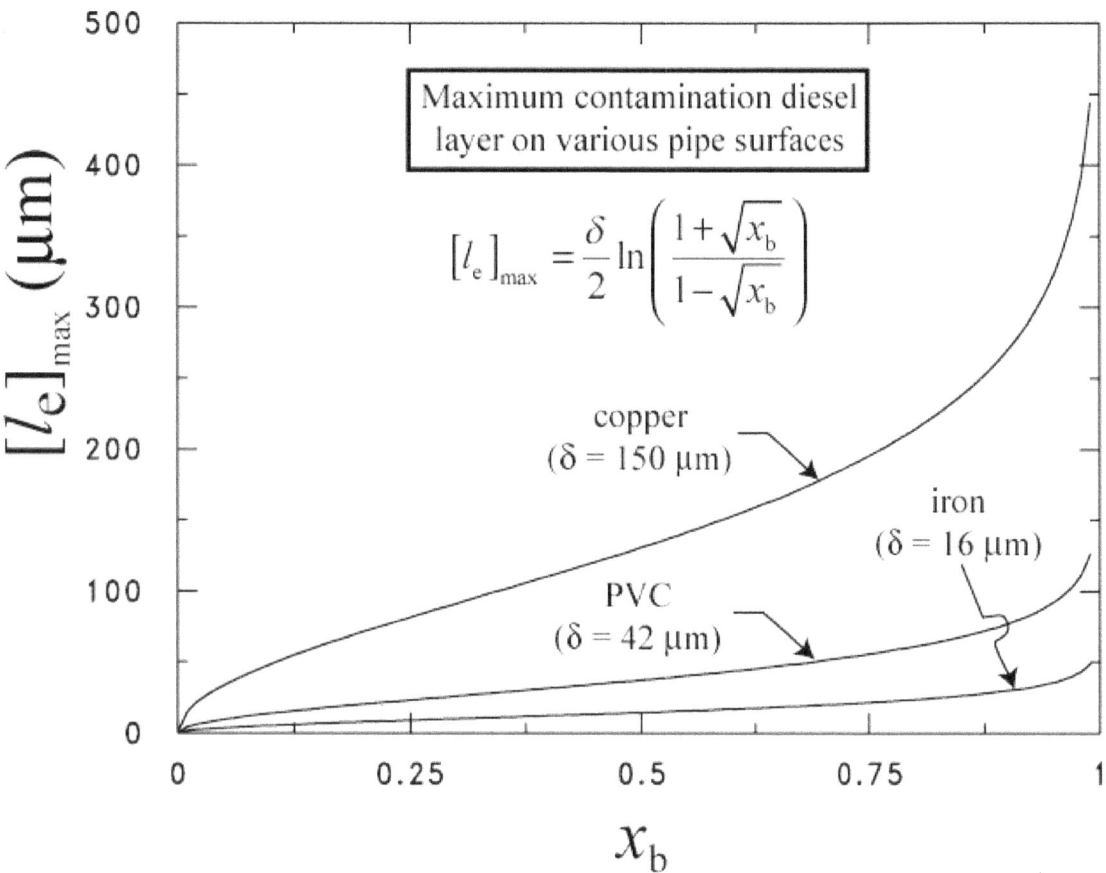

Fig. 5.1.2.2 Maximum contamination diesel layer on various pipe surfaces
as predicted by eq. (3.12)

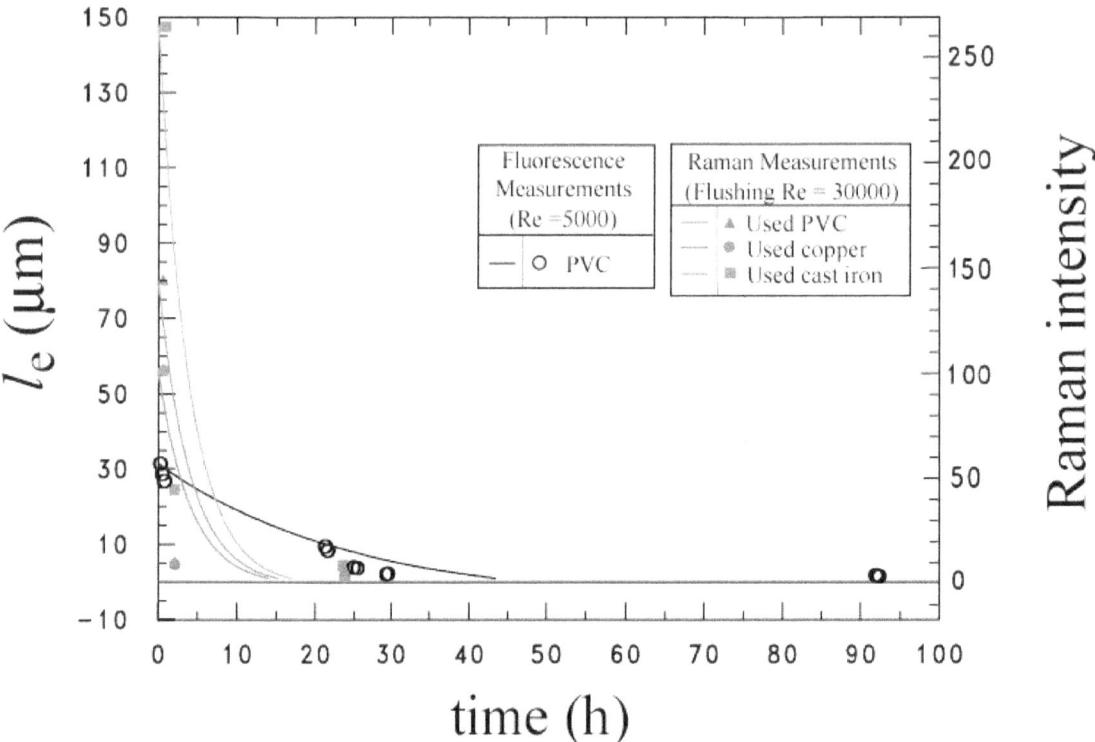

Fig. 5.1.2.3 Preliminary Validation of Semi-empirical Flushing Model

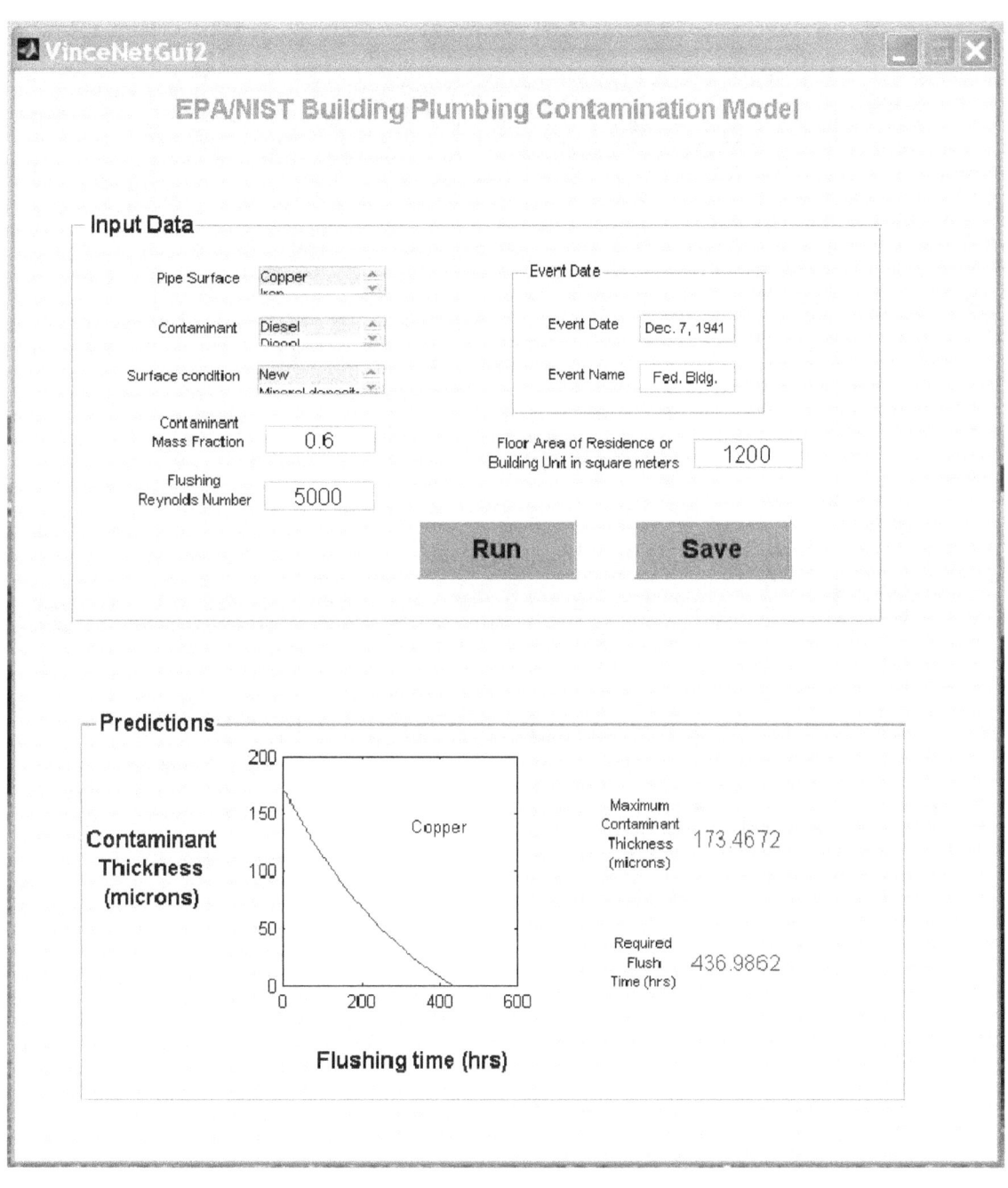

Fig. 5.1.4.1 Sample GUI for EPA/NIST Building Plumbing Contamination Model

143

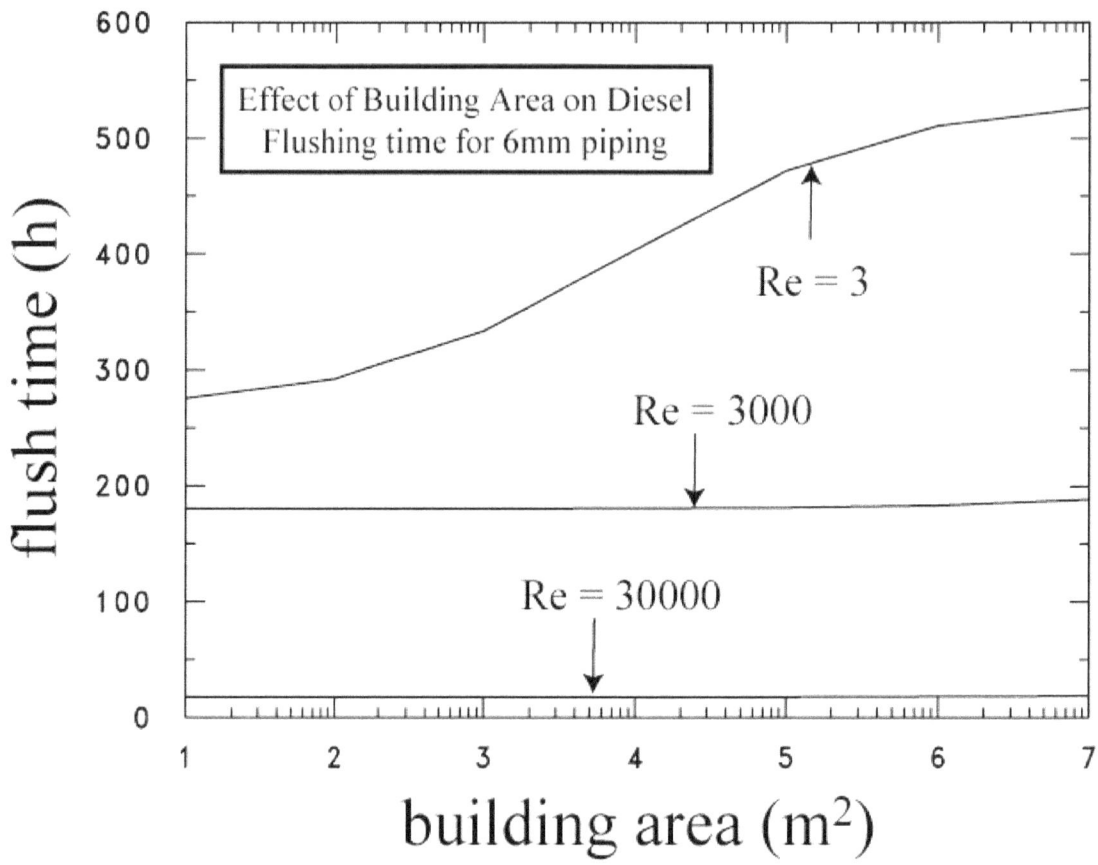

Fig. 5.1.4.2 Effect of Building Area on Diesel Flushing time for 6 mm piping

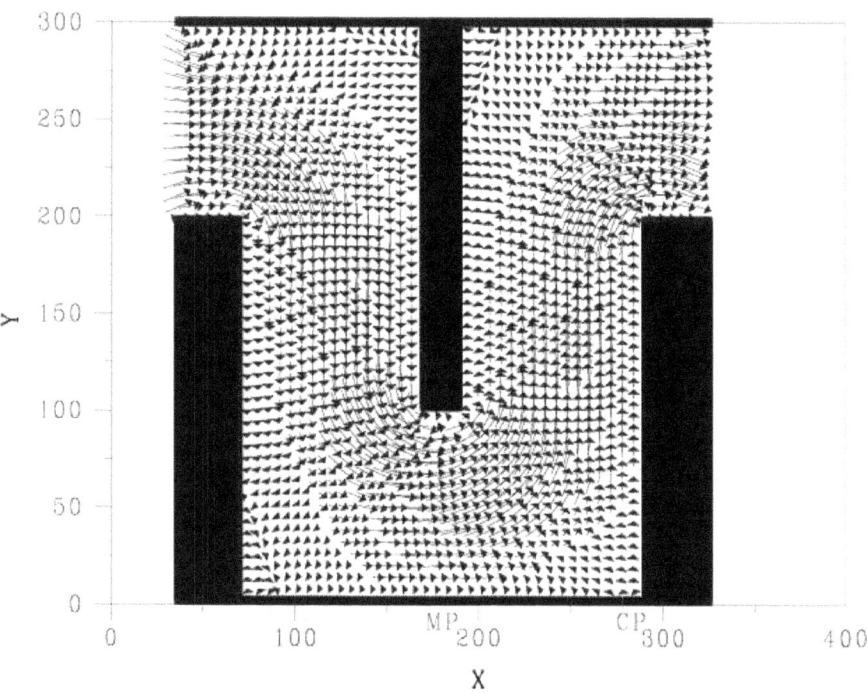

Fig. 5.2.1. Velocity fields, as indicated by the arrows, for a Reynolds number of about 30 in the U shaped pipe system. Contaminants were placed near the lower midway part of the flow path (MP) and at the bottom right hand corner (CP) of the pipe. X and Y represent the coordinates and length is in units of lattice spacing. Note that the simulation resolution is 6 times higher than that indicated by the velocity vectors.

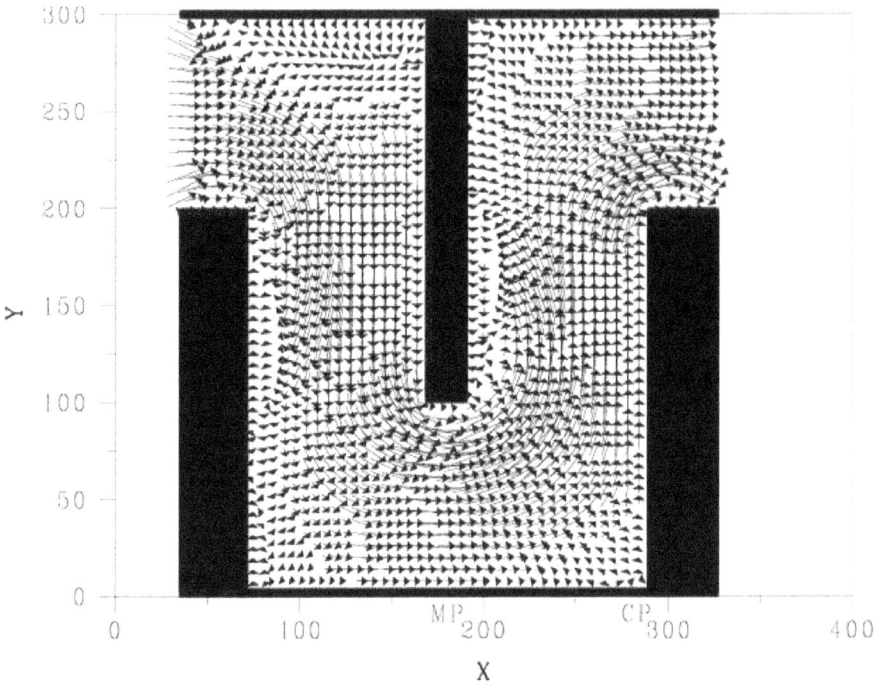

Fig. 5.2.2 Velocity fields for a Reynolds number of about 3000 for the U shaped pipe system. Note the regions of rotational flow near the corners and bends. Patches are located near MP and CP.

146

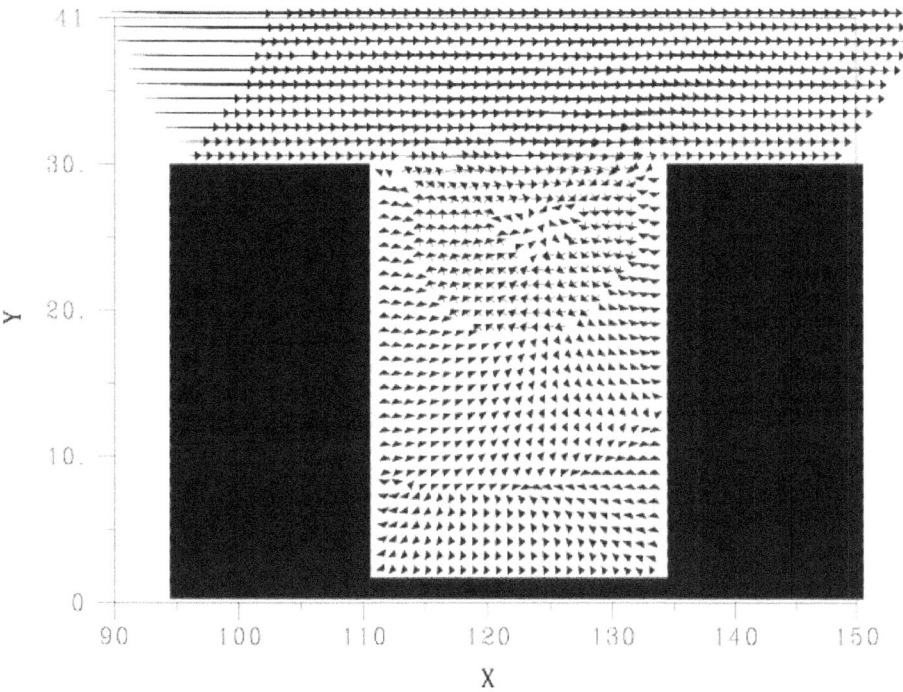

Fig. 5.2.3. **Flow near a cavity. Note the rotational flow near the opening of the cavity. A secondary rotation pattern developed near the bottom of the cavity. For this geometry, the contaminant was placed along the bottom of the cavity.**

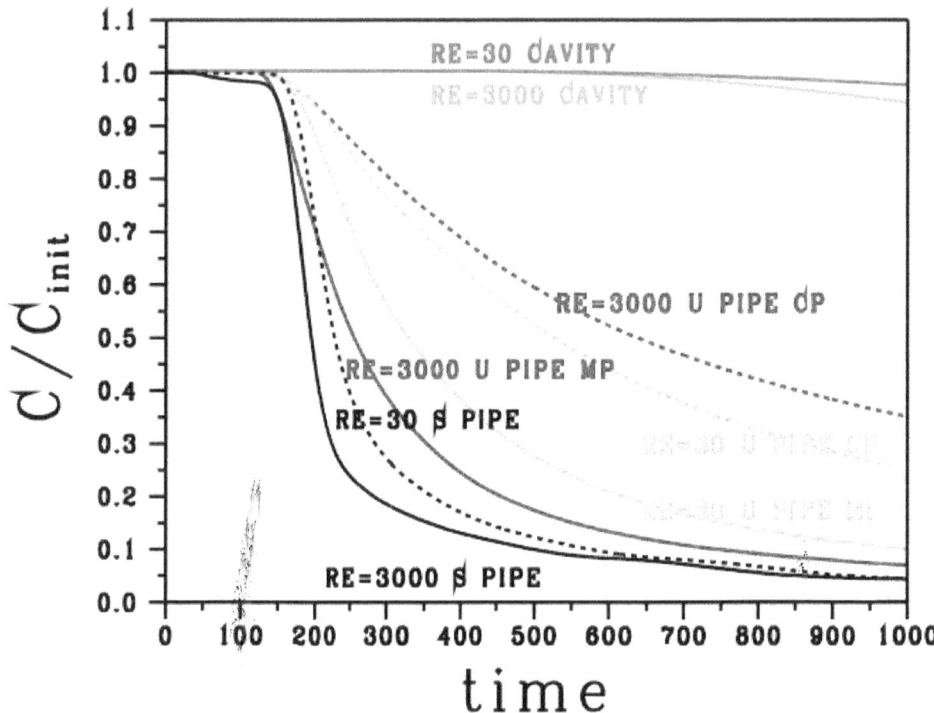

Fig. 5.2.4. Normalized pipe contamination, C/C$_{init}$. Time is given in relative units where 100 corresponds to the time it takes for the fluid to pass through the pipe system. The black lines (Re=30 dashed, Re=3000 solid) correspond to straight pipe flow (S), green and blue are for the U shaped pipe and the violet and purple for the cavity. The dashed (green or blue) lines correspond to case where the contaminant was placed in a corner (CP) and the solid lines all correspond to the contaminant placed midway (MP) in the U shaped pipe.

148

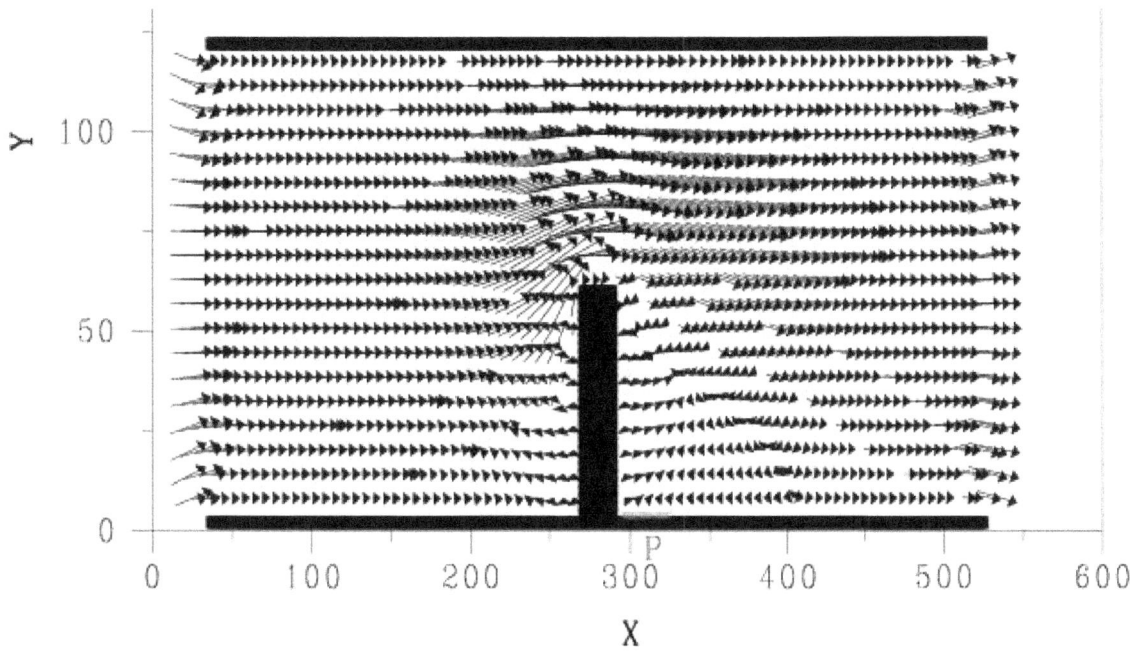

Fig. 5.2.5 Low Re number flow past a rectilinear obstruction.

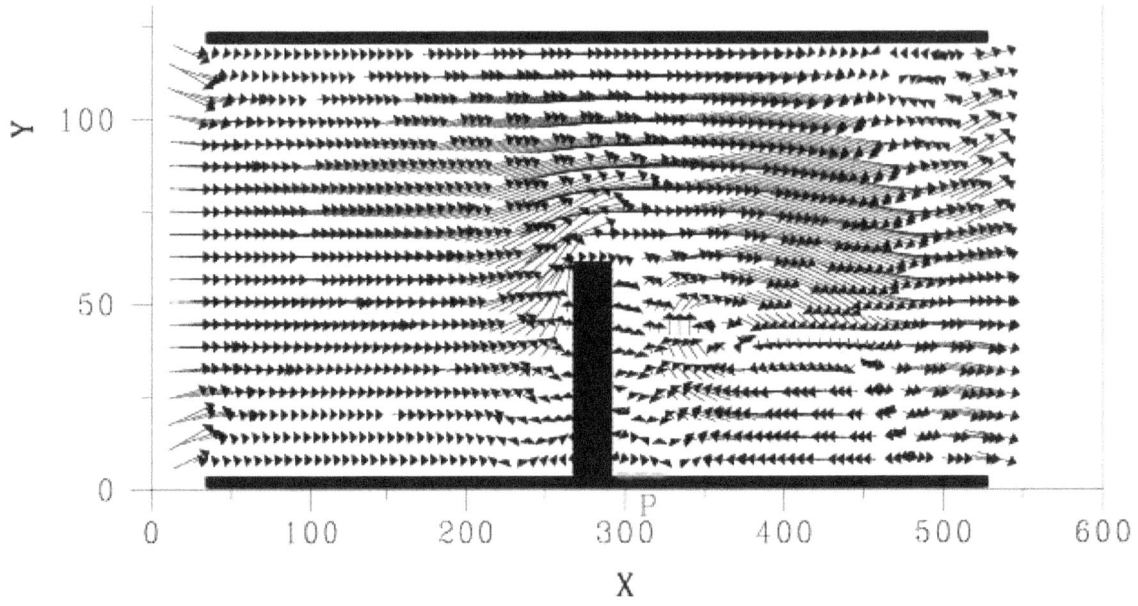

Fig. 5.2.6. High Re flow past a rectilinear obstruction. Note the well-defined vortex downstream.

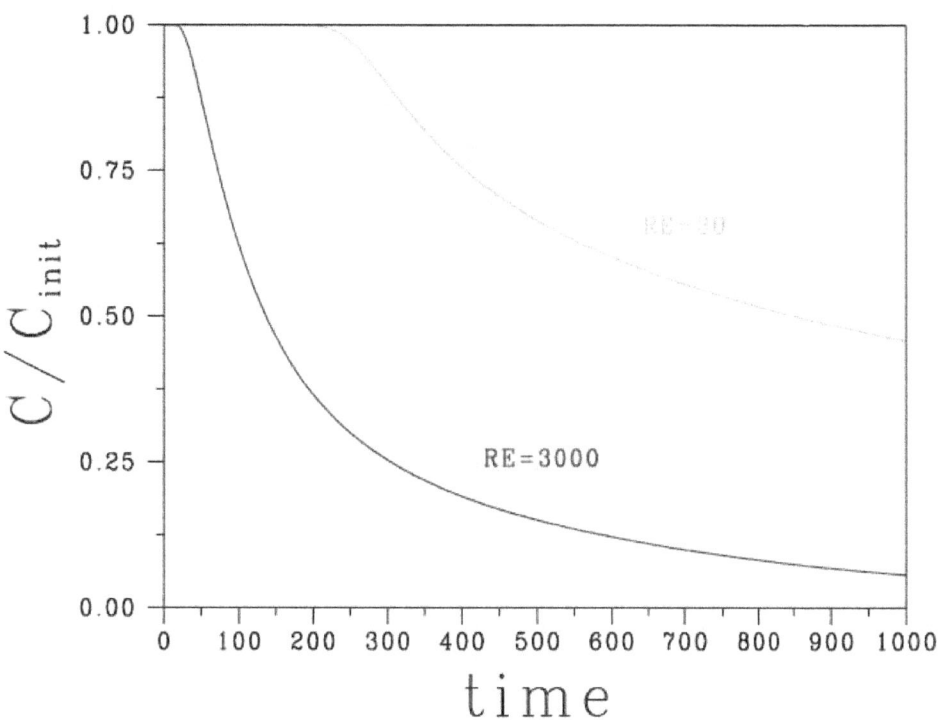

Fig. 5.2.7. Normalized total concentrations for the rectilinear cases.

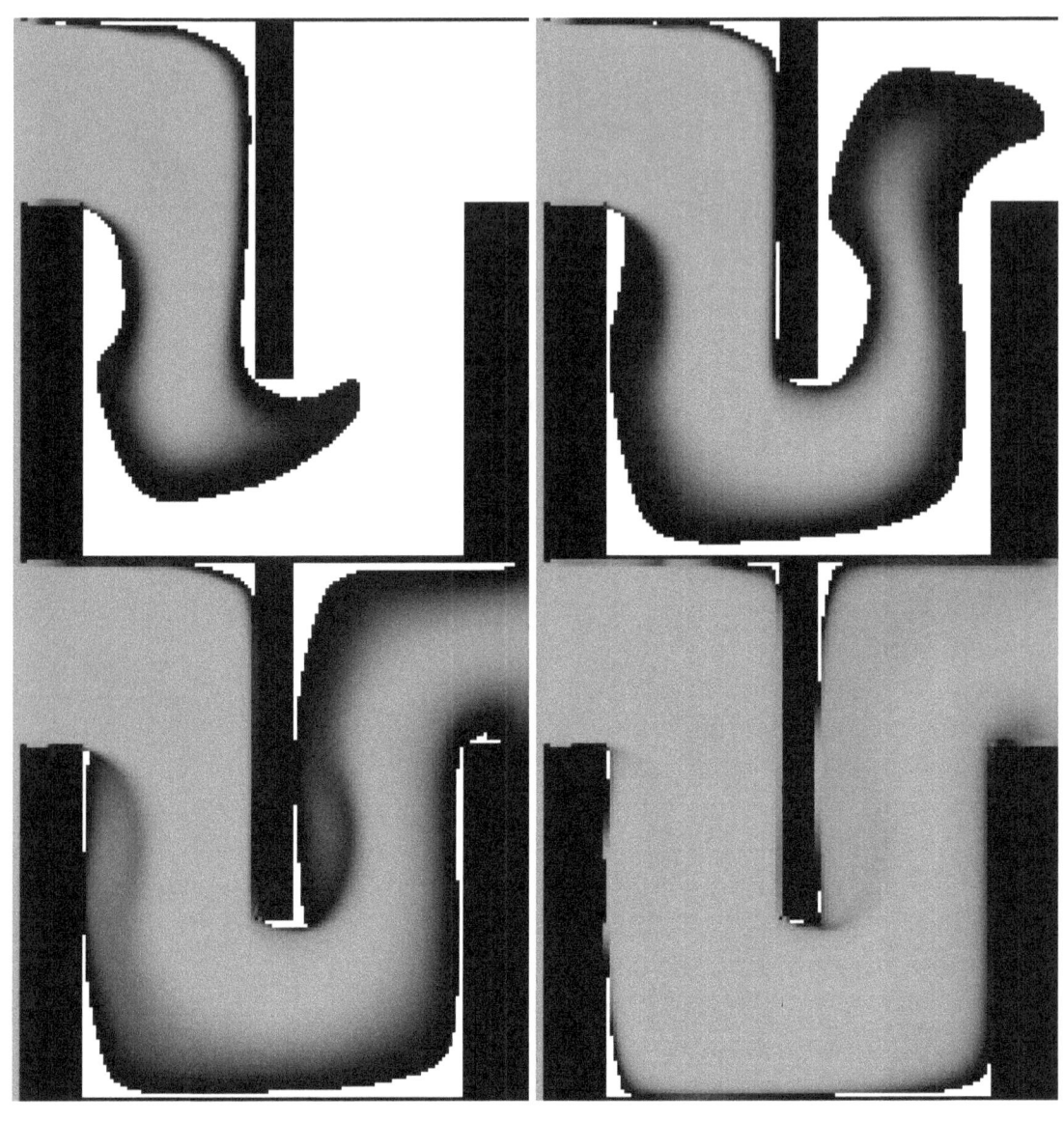

Fig. 5.2.8. Time sequence of ingress of contaminant. Note contaminant reaches the middle patch (MP) before it reaches the corner patch (CP).

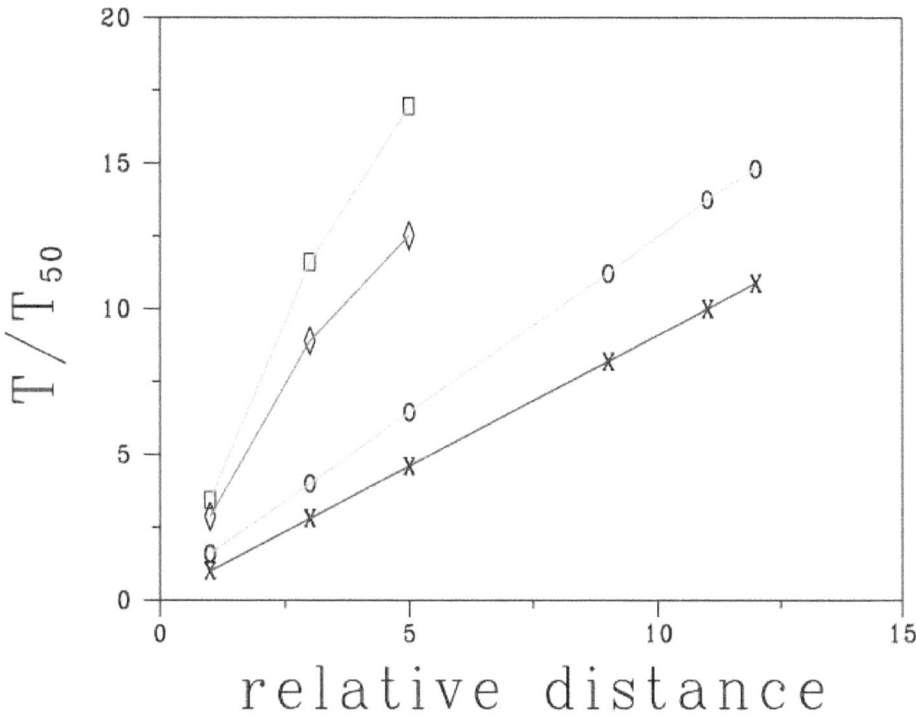

Fig. 5.2.9. The relative effect of pipe length on contaminant retention times. Time, T, was normalized to the time, T$_{50}$, it took the patch closest to the outlet to reach 50 % of its original value. The curves in blue, green, purple, and violet correspond to remaining contaminant levels of 50 %, 5 %, 1 %, and .5 % respectively.

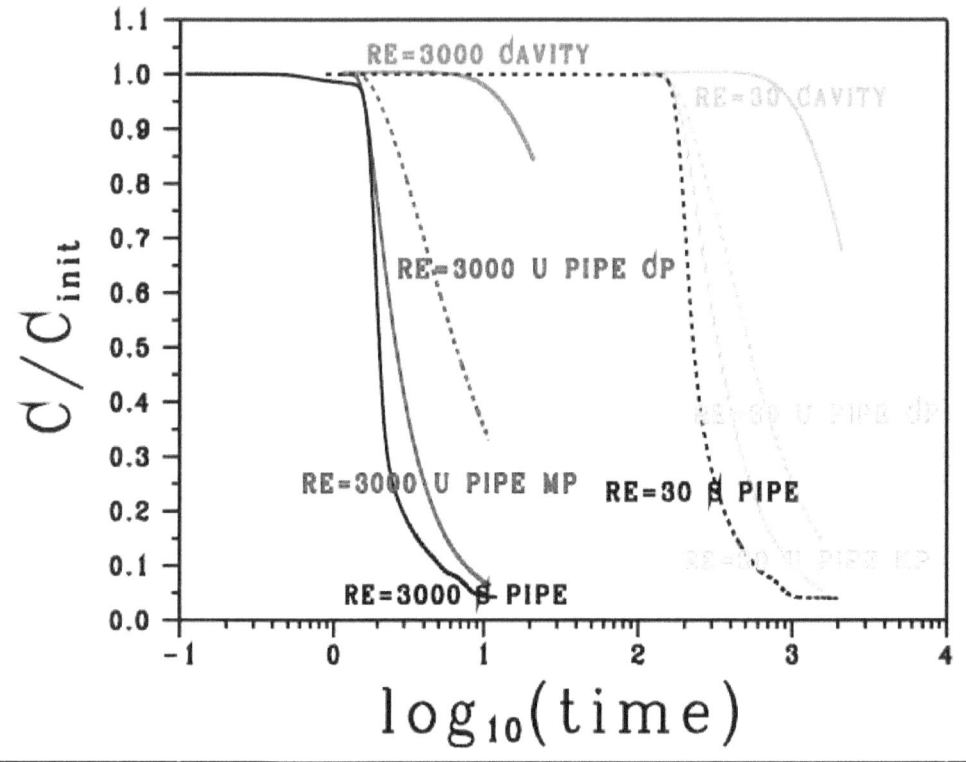

Fig. 5.2.10. Re-scaled total concentration adjusted for actual flow rates. Clearly, the higher Re flow removes the contaminant sooner for the cases studied.

154

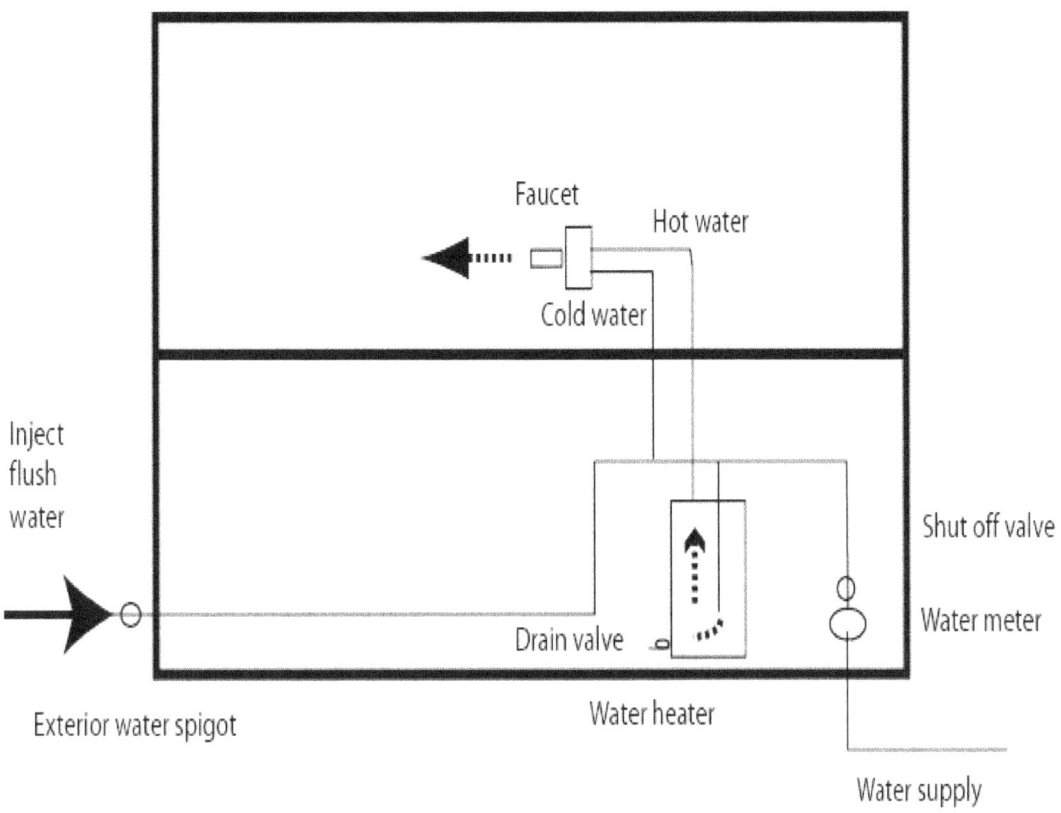

Fig. 6.3.1 Injecting flush water through an exterior water spigot or similar point will allow flushing of both hot and cold water lines and water heaters.

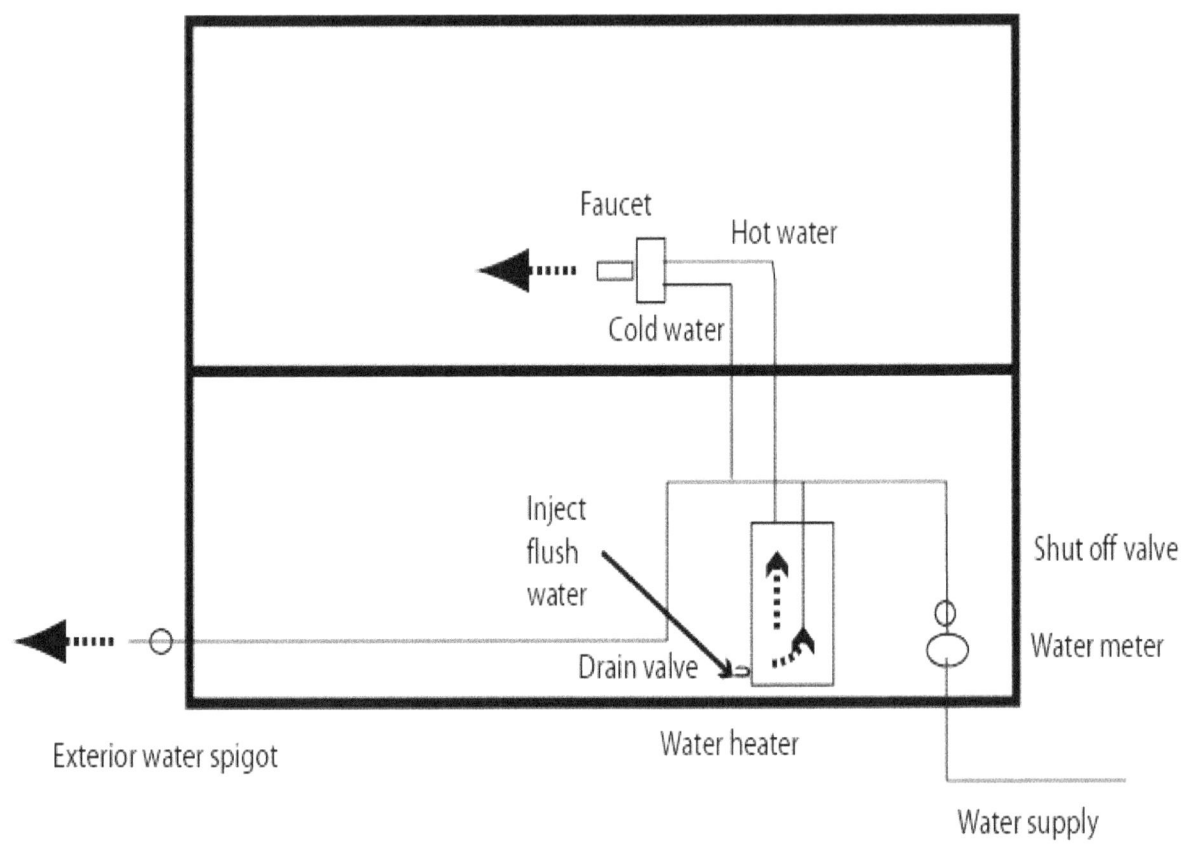

Fig. 6.3.2   Injecting flush water through the drain valve of a hot water heater can directly flush the tank and water lines.

www.ingramcontent.com/pod-product-compliance
Lightning Source LLC
Chambersburg PA
CBHW080250180526
45167CB00006B/2475